Introduction to the Biogeochemistry of Soils

This is the first process-based textbook on how soils form and function in biogeochemical cycles, offering a self-contained and integrated overview of the field as it now stands for advanced undergraduate and graduate students in soil science, environmental science, and the wider Earth sciences. The jargon-free approach quickly familiarizes students with the field's theoretical foundations before moving on to analyze chemical and other numerical data, building the necessary skills to develop questions and strategies for original research by the end of a single-semester course. The field-based framework equips students with the essential tools for accessing and interpreting the vast United States Department of Agriculture soil data set, allowing them to establish a working knowledge of the most important modern developments in soil research. This textbook is complete with numerous end-of-chapter questions, figures, and examples, and students will find it a multidisciplinary toolkit invaluable for their future careers.

Ronald Amundson has spent his career at Berkeley working to integrate and expand an appreciation of soils within the earth sciences. By utilizing principles of isotope geochemistry, he has developed isotopic tools for environmental and paleoclimate studies, and helped develop new methods of dating soils and landscapes. He has nearly two decades of research experience in the Atacama Desert of Chile, exploring the climate threshold between the biotic and abiotic regions on Earth, and how this can ultimately inform us about the history of our planetary neighbor, Mars. Amundson is an elected fellow of the Soil Science Society of America and the American Geophysical Union.

T0192878

Introduction to the Biogeochemistry of Soils

RONALD AMUNDSON

University of California, Berkeley

CAMBRIDGE
UNIVERSITY PRESS

University Printing House, Cambridge CB2 8BS, United Kingdom

One Liberty Plaza, 20th Floor, New York, NY 10006, USA

477 Williamstown Road, Port Melbourne, VIC 3207, Australia

314–321, 3rd Floor, Plot 3, Splendor Forum, Jasola District Centre,
New Delhi – 110025, India

79 Anson Road, #06–04/06, Singapore 079906

Cambridge University Press is part of the University of Cambridge.

It furthers the University's mission by disseminating knowledge in the pursuit of
education, learning, and research at the highest international levels of excellence.

www.cambridge.org
Information on this title: www.cambridge.org/9781108831260
DOI: 10.1017/9781108937795

First published 2021

A catalogue record for this publication is available from the British Library.

ISBN 978-1-108-83126-0 Hardback
ISBN 978-1-108-93275-2 Paperback

Additional resources for this publication at www.cambridge.org/amundson

In memory of my parents, Merle and Ethel, and my sister Susan.

And is it so hard to believe that souls might also travel those paths? . . . That great shuttles of souls might fly about, faded but audible if you listen closely enough?

Anthony Doerr – All the Light We Cannot See

Contents

Preface

Soil is now in the midst of an enormous increase in scientific recognition and interest due to its central importance to the challenging societal and global issues of climate change and food production. New soil biogeochemical theories and models in the past few decades, combined with easily accessible data sets, make understanding soil processes at the local to the global scale accessible not only to researchers but to students. However, much of the recent evolution in soil biogeochemical theory, or applications of models from other fields to soil biogeochemistry, is not available in a comprehensive manner in most books. One must largely access the primary peer-reviewed literature in order to learn about these methods and how they can be applied to specific questions. Thus, the purpose of this book is to serve as an introduction to recent developments in concepts and models of soil processes, and to the array of soil chemical and physical analyses that are now widely available. The book also applies the various concepts and tools to specific examples. Most importantly, the primary objective is to reveal how the reader might use this book as a starting point to address questions and problems of their own and advance the field through their own unique work.

Biogeochemistry, as the name indicates, spans biotic to abiotic chemical properties and processes. Here, the focus is largely on the soil solid phase and on the organic and inorganic compounds that comprise it and that undergo changes mediated by an array of complex mechanisms. Specifically, the focus is on soil processes occurring *in situ* on the landscape, with a view of soil as a functioning physical body exchanging matter and energy with its surroundings. This focus of observation then helps define and restrict the types of processes, and models to describe them, that are explored here.

1 Introduction to Soils

Certainly no clear line of demarcation has as yet been drawn between species and sub-species ... or again between sub-species and well-marked varieties, or between lesser varieties and individual species. These differences blend into each other in an insensible series; and a series impresses the mind with the idea of an actual passage.[1]

Charles Darwin, *Origin of Species*

1.1 Introduction

The soils that blanket most of the Earth's land surface are a membrane through which water, solids, and gases pass or with which they interact. These processes impact global climate, regional and global hydrology, and water chemistry, and the soils themselves are stores of largely unknown biological diversity that are only now being examined in an exploratory manner. Their importance to the functioning of the planet, and to human outcomes, makes it an exciting time to study them and their processes.

Pedology[2] is a natural science that concerns itself to a great extent with the biogeochemical processes that form and distribute soils across the globe. The word, and the science, originated during the scientific expansion of the nineteenth century,[3] a period of intellectual innovation that resulted in the development of geology and other branches of the natural sciences. Of course, soil had been examined by farmers, early scientists, and philosophers for thousands of years prior to the 1800s but primarily as a medium for plant growth and other agricultural or industrial uses. It was the epiphany of Russian and American scientists[4] that soil is a three-dimensional component of the Earth's surface, and that it is predictably distributed in relation to certain environmental and geological factors, that elevated the study of soil into a bona fide branch of science.

Soil is the object of study in pedology. While the science of pedology has a broadly accepted definition, there is no precise definition for soil as a three-dimensional object, nor is there likely ever to be one. The reason for this paradox is that soil is part of a continuum of materials at the Earth's surface.[5] At a soil's base, the exact line of demarcation between "soil" and "nonsoil" will continue to elude general agreement because the chemical and physical changes induced by pedogenesis disappear gradually, commonly over great vertical distances. Similarly, an identical problem confronts anyone attempting to delineate the boundary between one soil "type" and another. Soil properties, such as horizons, commonly change gradually and continuously in a horizontal direction, which leads to a

$$\frac{\partial S}{\partial t}$$

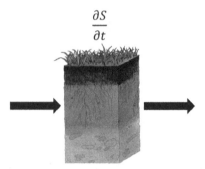

Figure 1.1 Schematic view of a soil, an open chemical system. Mass and energy enter and exit, causing changes to the state of the system over time.

view that the Earth possesses an infinite variety of soils.[6] A science with a "poorly" defined object of study is not peculiar to pedology. As Charles Darwin recognized in the *Origin of Species*,[7] an exact definition of a species, or a living being,[8] may elude a biologist, yet this has in no way hindered profound advances in our understanding of life.

To contend with the soil continuum, scientists partition it, albeit arbitrarily, into systems that suit the need of the investigator. Systems are human constructs that confine our focus and allow us to develop quantitative tools to evaluate a portion of the soil continuum. These systems are open to their surroundings and allow the passage and transfer of energy and matter (Figure 1.1). The properties inside a physical system vary in response to certain sets of factors: the initial state, the surrounding environment, and time.[9] The formulation of soil systems, and their dependence on sets of factors, constitutes the paradigm of pedology that was first formulated in general terms by the visionary Russian scientist V. V. Dochuchaev[10] and later cast into a more formal theoretical framework by the American Hans Jenny.[11] These concepts about soils also have a direct bearing on our understanding of ecosystems (because soil is a key part of terrestrial ecosystems). This theory ultimately provides a definition of soil that, while not precise in identifying its boundaries, is consistent with this theoretical framework: *"soil is those portions of the earth's crust whose properties vary with soil forming factors."*[12]

This definition of soil recognizes it as a natural component of nature. At a finer scale (i.e. the nature of what is within the system), soils have been described as "multicomponent, multiphase, open systems that sustain a myriad of interconnected chemical reactions, including those involving the soil biota."[13] Later, the chemical and mineralogical makeup of the solid phase will be examined. Likewise, the enormous importance of biota for soils will be discussed under several topics, particularly in terms of their effect on the C cycle. Finally, the interaction between the inorganic and organic components will be examined through the process of weathering, and its chemical and physical implications for soil characteristics will be evaluated. Throughout, however, the focus will be on properties and processes observable *in situ* in nature, and thus, some of the microscopic and molecular details will be bypassed in order to begin to appreciate the more macroscopic spatial patterns found in nature.

1.2 Factors Defining the Soil System

The soil system, and the factors that define it, will now be elucidated to understand how soils form and how they can be quantitatively examined. Hans Jenny applied principles from the physical sciences to the study of soil systems and soil formation. Briefly, Jenny recognized that soil systems (or if the above-ground flora and fauna are also considered, ecosystems[14]) exchange mass and energy with their surroundings and that their properties can be defined by a limited set of independent variables. Based on comparisons with the physical sciences, Jenny's state factor model of soil formation states that:

$$\underbrace{Soil/Ecosystems}_{\substack{dependent \\ variables}}$$

$$= f(\underbrace{initial\ state\ of\ system, surrounding\ environment, elapsed\ time}_{\substack{independent \\ variables}}) \qquad (1.1)$$

where the terms on the right-hand side of the equation constitute independent variables, which, when combined with the indeterminate function f, define the state of the system. The somewhat generic variables can be further refined, in the case of soil systems, as a result of innovations by Dokuchaev and others. Their work identified a set of more specific environmental factors that encompass the generic factors listed earlier:

$$Soils(S)/Ecosystems(E)$$

$$= f(\underbrace{climate(cl), organisms(o),}_{\substack{surrounding \\ environment}} \underbrace{topography(r), parent\ material(p),}_{\substack{initial\ state \\ of\ soil}} time(t), ...) \qquad (1.2)$$

The variables on the right side of Eq. (1.2), the so-called "soil-forming factors,"[15] have these important characteristics: (1) they are independent of the system being studied[16] and (2) in many parts of the Earth, the state factors vary independently of each other (though, of course, not always). As a result, through judicious site (system) selection, the influence of a single factor can be observed and quantified in nature.

Based on the characteristics of the factors discussed earlier, Jenny cast Eq. (1.2) into differential form, opening some important conceptual and quantitative avenues:

$$dS = \frac{\partial S}{\partial cl}dcl + \frac{\partial S}{\partial o}do + \frac{\partial S}{\partial r}dr + \frac{\partial S}{\partial p}dp + \frac{\partial S}{\partial t}dt \qquad (1.3)$$

where the quotients on the right side are the partial derivatives of the functions relating any soil property to that independent variable (e.g. $\frac{\partial S}{\partial factor} = f'(factor)$).

The goal of many biogeochemical studies of soil is to empirically determine the form of the function f(factor), since these functions, as Jenny pointed out, cannot be reliably determined on strictly theoretical grounds. In order to determine the nature of these functions in the field, a series of sites must be selected in which only the variable of interest is allowed to change, and other variables are held constant. Equation (1.3) provides two ways in which other variables may be "held constant." First, the absolute range in a factor (∂ factor) can be held at 0, eliminating confounding impacts from this variable (e.g. a series of sites, all the same age but in different climates, could be used to examine the effect of climate on a soil property). Second, a variable may be "held constant" in a field study if its partial derivative $(\frac{\partial S}{\partial factor}) \cong 0$. For example, in the case of soil C, C is strongly dependent on time, but it reaches steady state in approximately 10^3 to 10^4 years. Thus, if the effect of another factor is of interest (e.g. climate), a series of sites in differing climates and of differing ages (as long as the sites are at steady state) may be used to examine the effect of climate.

A second aspect of the differential form of the state factor equation (Eq. (1.3)) is that it is observationally possible (though certainly a considerable effort) to determine the relative importance of individual factors for a given soil property through a series of single-variable studies, so that the absolute and relative importance of $(\frac{\partial S}{\partial factor})$ for all state factors on a soil property may be determined. To date, this has been only partially achieved, but with growing data sets obtained using a state factor concept (combined with multivariate statistical analyses and emerging advances in data science), it may be possible to begin to conduct and construct such comparisons if the data have been properly collected. The current "Critical Zone" research program of the National Science Foundation consists of a series of field sites specifically selected to examine the effect of an array of the state factor variables (Figure 1.2).

One criticism of Eq. (1.3) that has been raised by a few is that it has never been solved. This apparently springs from the fact that there is indeed no single numerical

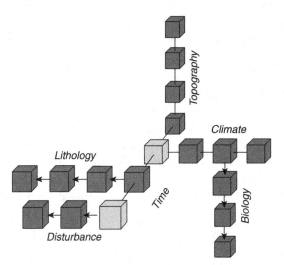

Figure 1.2   Conceptual design of the Critical Zone Observatory network. A number of sites, in varying state factor space, will allow the development of quantitative functions. From S. L. Brantley et al., *Frontiers in Exploration of the Critical Zone.* Report of a workshop sponsored by the National Science Foundation, October 24–26 (2005), Newark, DE, 30p.

Table 1.1 The major factors of soil and ecosystem formation, and a brief outline of their characteristics

State Factor	Definition and Characteristics	Common Parameterization
Climate	Regional climate	Mean annual temperature (^{0}C), mean annual precipitation (mm)
Organisms	Potential biotic flux into system	Very few studies
Topography	Slope, aspect, and landscape configuration	slope (%), curvature (L^{-1})
Parent Material	Chemical and physical characteristics of soil system at t = 0	Chemical/mineralogical content
Time	Elapsed time since system was formed or rejuvenated	yr^{-1} or Ky^{-1}
Humans	A special biotic factor due to magnitude of human alteration of earth and humans' possession of variable cultural practices and attitudes that alter landscapes	Very few studies

model to describe the effect of all factors on any given soil property, or an entire soil, for the whole of the planet. Such a goal still is, and will likely remain, elusive. However, this view does not account for the growing number of solutions for specific locations and for specific properties – especially for climatically and societally important soil properties such as carbon and nitrogen. A goal of this book is to illuminate how one can analyze and use data effectively, and the following text will include some examples of solutions to Eq. (1.3) and what they mean for understanding soils in global processes.

Table 1.1 provides a brief definition of the state factors of soil formation and some ways in which they have been numerically or qualitatively represented. A field study designed to observe the influence of one state factor on soil properties or processes is referred to as a sequence: for example, a series of sites that have similar state factor values except climate is referred to as a climosequence. Similar sequences can be, and have been, established to examine the effect of other state factors on soils. A review of soil state factor studies was presented by Birkeland.[17] A set of papers discussing the impact of Jenny's state factor model on advances in pedology, geology, ecology, and related sciences is presented in Amundson et al.[18]

The state factor approach to studying soil formation is an effective quantitative means of linking soil properties to important variables[19] and is a way of designing experiments. There is a large and growing assemblage of soil data for the USA and the globe. As data acquisition and analytical tools improve, these data are being more widely examined for insights into factorial relationships. A challenge to uncovering clear principles, such as dependence of a property on climate, is that data sets may also have enormous ranges of lithologies, topographic positions, soil ages, land use, etc. that obscure the signal of, for example, climate.

One of the most commonly investigated state factors is climate. Climate has a first-order control on many of the processes that will be examined in this book, particularly C cycling, chemical weathering, and hillslope soil processes. In a number of instances, using large data

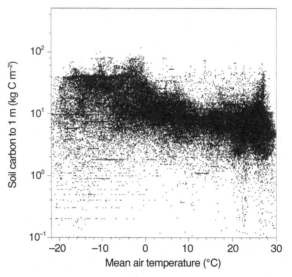

Figure 1.3 Scatterplot of soil C as a function of mean annual temperature, as compiled from large global soil data sets. C. D. Koven et al., Higher climatological temperature sensitivity of soil carbon in cold than warm climates, *Nature Climate Change*, doi: 10.1038/NCLIMATE3421 (2017).

sets spanning significant climate gradients, where other variables were not held constant, the climate impact still emerges, though with more variation due to the secondary impacts of the remaining state factors. For example, recently Koven et al.[20] used two large global soil databases to explore the temperature sensitivity of soil C storage to temperature (Figure 1.3). The plot shows that as temperature increases, soil C decreases, because, as will be discussed later, decomposition rates correspondingly increase. The results are statistically significant, but the "noise" in the trends reflects variations in rainfall, slope position, etc. By carefully designing a study to control for variations in other state factors, one can more clearly observe the effect of the variable of interest.

An ideal location to design natural experiments is the midcontinent of North America. There, rainfall and temperature vary systematically with latitude (T) and longitude (P). Additionally, large areas have similar parent materials and ages, so that the effect of climate on specific soil properties of interest can be determined. Jenny[21] recognized this during his first visit to the area as part of the International Congress of Soil Science field tour in 1927, and immediately upon his appointment as assistant professor at the University of Missouri, he began assembling climate data and previously published soil N data for the USA and Canada to test the hypothesis (and the relations) between climate and N in soils. Jenny's research, and that of others who followed, illustrates two important ideas emphasized throughout this book: (1) the power of state factor gradients to understand patterns and rates of soil processes, and (2) the fact that data to address many questions already exist in libraries and now, of course, more frequently at one's fingertips on the Web. Jenny's award-winning original research was done with climate data and soil analyses that had been published at the time, but now the amount of information completely dwarfs that which was available 100 years ago.

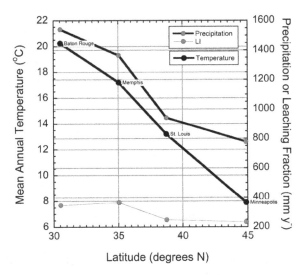

Annual temperature, precipitation, and calculated water balance vs. latitude along the Mississippi River corridor. Data from NOAA and WorldClimate.org.

One of the gradients that Jenny explored is a N–S transect from Wisconsin to Mississippi (which he called "humid prairie soils – silt loam"). Subsequent research, which we will begin to discuss here and return to later in the book, has followed to focus on the chemical weathering in soils along this same transect. The soil transect follows the loess-covered uplands just east of the Mississippi River corridor, loosely traced along its entirety by US Highway 61.[22] Climatically, temperature declines to the north, as does precipitation. Jenny recognized that the water balance (he used a "humidity index" = precipitation/saturation deficit of air) was largely similar along the gradient, and thus, the climatically driving variable can be considered to be temperature. Since Jenny's work, other metrics for water balance have been developed, and in Figure 1.4, the *leaching index*, or the total of monthly precipitation that exceeds evapotranspiration (using the Thornthwaite model), is shown for four major cities along the transect, confirming that available soil water for leaching is very similar, but the temperatures vary greatly.

Figure 1.5 shows Jenny's soil N data, for the upper 15 cm, for uncultivated sites from Wisconsin to Mississippi. The results show that N (or if one multiplies by approximately 10 (C/N ratio), the C%) increases with decreasing temperature. Later, we will see that this is a balance of N and C inputs and losses controlled by climate. In the midcontinent, transects can also be developed along a roughly E–W direction, where temperature is constant and rainfall changes. Finally, by taking somewhat NW to SE transects in some locations, precipitation can be held constant while temperature changes.[23] Jenny and others have used these opportunities to disentangle the effects of temperature and precipitation on soil C storage.[24]

More recently, geoscientists have quantified the total soil chemical changes along the Mississippi corridor gradient.[25] As we will learn, most of the mass of soils consists of primary silicate minerals, which undergo slow rates of chemical alteration to release biogeochemically

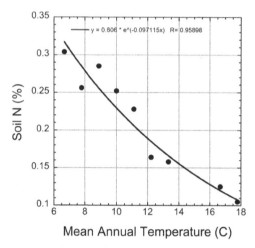

Figure 1.5
Soil N (upper 30 cm) vs. mean annual temperature. Data from H. Jenny, A study of the influence of climate upon the nitrogen and organic matter content of the soil, Missouri Agricultural Experiment Station Bulletin, 152 (1930).

important elements, such as Ca, which are an important part of many biochemical mechanisms. Also, it will be shown that it is more accurate to compare the concentration of an element that can be chemically released and removed with that of an element that is insensitive to removal (Ti). The authors calculated the CaO/TiO_2 ratio for the entire soil profile at 22 sites along this gradient (Figure 1.6). The results show that from the northernmost to the southernmost site, the ratios of CaO to TiO_2 declined from 1.7 to 0.4. As discussed, the amount of precipitation that falls and moves through the soils each year is roughly the same at all sites, due to higher evapotranspiration rates with increasing temperatures. However, the temperature, duration of warm seasons, and associated biological activity increase greatly with increasing temperatures. These changes all facilitate chemical weathering, which is reliant on (1) temperature and (2) acidity, in addition to water flux. While it might be tempting to compare how both N and Ca/Ti change with temperature by examining the first derivative of the exponential models' fit through the data, this is not an accurate analysis, in that the soil N is roughly at steady state (a constant value), while the Ca/Ti ratios continue to change (albeit slowly) with time. The original climate sequence that Jenny exploited continues to be used for other environmentally relevant questions that are important today.

1.3 Summary

Pedology is a science that is commonly observational (e.g. field-based observations of long-term "natural experiments") rather than experimental in nature, concerned with the origin and functioning of soils in place on the landscape. While experimentation, and concepts from experimental sciences, informs our understanding of soils, the vast expanse of time

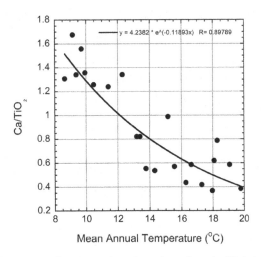

The Ca to Ti ratios of soil profiles along a latitude gradient along the Mississippi River. Data from D. R. Muhs et al., Impact of climate and parent material on chemical weathering in loess-derived soils of the Mississippi River Valley, Soil Science Society of America Journal, 65, 1761–1777 (2001).

required to form soils requires pedologists to develop tools, concepts, and modes of scientific enquiry for problems not always confronted in the lab.

Pedology, like soil itself, is part of an intellectual continuum. There is no clear demarcation between pedology and related branches of the natural sciences, particularly geology and ecology. In many ways, pedology has as much in common with these sister sciences as with the "traditional" branches of the soil sciences such as physics, chemistry, etc., which are primarily experimentally driven.

The conceptualization of soil as a physical system, controlled by an array of factors that are independent of the system itself, is a powerful tool that allows one to develop and explore the mathematical relationships between these factors and soil properties of interest. These relations can then be useful, alone or with other relations, in ultimately understanding how soils function and are distributed.

1.4 Activities

1.4.1 Soil in Current Events

To appreciate the enormous importance of soil to society, conduct a web search of "soil in the news." For example, the *New York Times* presently maintains a compilation of recent soil-related articles under the heading "Soil." Pick three news items, read them, and provide a paragraph discussion of how the state factor theory may be relevant to the issues. In 2020, the following three issues are chosen as examples of "soil in the news."

China's Rover Finds Layers of Surprise Under Moon's Far Side

This article reports on the results of ground-penetrating radar analysis of the *regolith* in a crater on the far side of the moon (look up, and define, the concept of regolith). The researchers found over 40 m of regolith, or soil, over bedrock. This soil has been formed over the eons by meteor impact and subsequent degradation of the moon's surface.

What are the values of common factors of soil formation for the moon, and, most importantly, what *additional* state factors might be important be in this unique planetary setting?

"Earthworm Dilemma" Has Climate Scientists Racing to Keep Up

This news article addresses the impact of European earthworms that have been introduced to the northern latitudes of North America. Some of these species facilitate the degradation of litter layers, release C to the atmosphere, and impact biodiversity.

From a state factor perspective, what factor has been impacted by the introduction of worms, and how can the state factor model be used to design observational experiments to study and quantify the changes over time?

To Combat Climate Change, Start From the Ground Up (With Dirt)

In this graphic essay, the author repeats many statements and sentiments made about soil, and our management of soil, as a means of mitigating and adapting to climate change. These heartfelt essays, by nonscientists, may sometimes misstate or overplay certain issues. For example, do some online research to determine the accuracy or basis of the following statements:

1. Soil stores 20× its weight in water (search topics like soil water-holding capacity)
2. It keeps carbon out of the atmosphere (search topics such as natural negative carbon emissions)
3. "Dirt is dead, soil is alive" (how does this compare with our definition of soil?)

1.4.2 Solving the State Factor Equation

In this chapter, the generic state factor model and equation were introduced. The real excitement, as a scientist, is to uncover quantitative relationships, insert them into the equation, and create models of practical significance. Here, we explore this, solving the same equation that Jenny first solved nearly a century ago.[26]

Soil scientists had long recognized that soil organic matter (which also contains most of the soil nitrogen) tends to decrease with increasing temperature and increase with increasing rainfall. However, no mathematical relationships had been derived. Jenny assembled hundreds of published, and new, soil analyses (N% in the surface horizon) and assembled the associated climatic information. Jenny recognized that the climate factor has at least two major components: temperature and moisture. In several papers between 1928 and 1930, Jenny pointed out that characterizing the temperature of a location is a complex function of season and other factors. He settled on using the parameter *mean annual temperature* largely because it was the most readily available metric for most weather stations. Many scientists still use this metric, though temperature can be defined or quantified in other ways. As for moisture, Jenny (as do many scientists today) recognized that total rainfall is likely less important to biological processes than is the *available rainfall*, which is the amount left after evapotranspiration. Jenny, using the data available, settled on using the *N.S. Quotient*, which is the value obtained by dividing precipitation by the absolute saturation deficit in the air.

With all this data, Jenny set out to solve:

$$N = f(cl)_{o.r,p,t,..}, or \quad N = f(T,H)_{o,r,p,t,...} \tag{1.4}$$

where T = MAT (°C) and H = NSQ.

While Jenny had not yet formalized the entire state factor equation at that point, he discussed effects of other factors and chose sites that were to some degree insensitive or constrained the effect of factors other than climate. The differential form of Eq. (1.4) is:

$$dN = (dN/dT)_H dT + (dN/dH)_T dH \tag{1.5}$$

which describes the sensitivity of N to changes in T and H via the (undefined) functions represented by the terms in parentheses. In physical chemistry, Eq. (1.5) is known as an "equation of state" (where the independent variables T and H define the soil N system), which is what inspired Jenny to call the "soil forming factors" the "state factors."

Equation (1.5) is a partial differential equation representing a multivariable function. If the partial derivatives (e.g. the functions that replace the terms on the right side of the equation) – which relate the manner in which N varies with respect to a state factor, with the others held constant – are known, then they can be combined to create a more complex equation after integration.

In order to put numerical relations between T, H, and N into Eq. (1.5), one has to first determine the nature of the function that relates the independent to the dependent variables. This can be started by plotting the data in an X, Y format and considering the shape of the relationships. Jenny did this, and also relied on an *a priori* understanding of how biology responds to temperature and how vegetation responds to rainfall. Jenny, relying on work he cited in his paper, used a variant of the van't Hoff relationship (or the related Arrhenius equation), recognizing that the rate of a biological process is exponentially related to temperature. As for water, Jenny used the concept that there are diminishing returns in productivity with increasing rainfall, a process described by the formula of Mitscherlich. These two models have the following forms:

$$N_H = C_1 e^{-k1T} \tag{1.6}$$

$$N_T = A(1 - e^{-k2H}) \tag{1.7}$$

where C_1 (N%), A (N%), k_1 (T^{-1}), and k_2 (mm^{-1}) are constants.

Today, many scientists use statistical software packages to find the best model that matches the equation. However, the approach here (taken by Jenny as well as others) is built on more general principles of how soil processes should vary with climate, so we follow that here as well.

In order to proceed, the first derivatives of Eq. (1.6) and (1.7) are required. The derivatives are:

$$\partial N/dT = (-C_1 k_1 e^{-kT}) = (and, after\ substitution\ of\ Eq.\ 4): -k_1 N \qquad (1.8)$$

$$\partial N/dH = (Ak_2 e^{-k2H}) = (which\ after\ substitution\ and\ rearrangement)$$
$$: [(Nk_2 e^{-k2H})/(1 - e^{-k2H})] \qquad (1.9)$$

Inserting the partial differentials Eq. (1.8) and (1.9) (e.g. the partial effects of both T and H on N) into the differential Eq. (1.5) gives:

$$dN = (-K_1 N)dT + [(NK_2 e^{-KH})/(1 - e^{-KH})]dH$$

By dividing by N (e.g. separating variables), one can integrate the equation to arrive at the solution for Eq. (1.4):

$$N = C_2 e^{-k1T}(1 - e^{-k2H}) \qquad (1.10)$$

Table 1.2 shows data that Jenny compiled (and averaged) for the USA.

Enter the data in Excel. Use the Solver function to fit the parameters (C_2, k_1, k_2) in Eq. (1.10), and list the results in a table. How close are your parameter values to those of Jenny ($C_3 = 0.55$; $k_1 = 0.08$; $k_2 = 0.005$)?

Add your values of the coefficients to Eq. (1.10). Create a matrix table in Excel of T and H, with the ranges of T from 0 to 22.5 C and H from 50 to 400 mm. Using Eq. (1.10), calculate N in all the combinations, and using the "surface" plot graphing features in Excel, create a "3D" N plot of N vs. MAT and H. Provide the plot, and discuss the general features of the plot and where one would expect to find soils high in N and low in N.

1.4.3 Sensitivity of Soil Properties to a State Factor

Various soil properties respond to state factors at differing rates, which underscores or reflects the sensitivity of the processes that produce them to the factor being examined. In a very well-designed study of the effects of rainfall (all other state factors being largely constant), Jenny and Leonard (1934)[27] examined the effect of rainfall on various soil properties. The rainfall across this transect of the central Plains ranged from 40 to 100 cm y^{-1}. The authors found that the following relations best described the trends for these soil properties (other properties were also examined):

$$N\% = 0.00258(MAP) - 0.0230$$
$$Clay\% = 0.360\ (MAP) + 1.330$$
$$pH = 9.516 - 0.043\ (MAP)$$

where MAP = mean annual precipitation in cm y^{-1}.

MAT	NSQ	N%
Table 1.2 Data compiled by Hans Jenny on mean annual temperature, NS quotient and percentage of nitrogen		
0	350	0.47
2.2	200	0.29
2.2	380	0.34
4.4	220	0.24
4.4	350	0.3
5.6	380	0.27
6.7	350	0.27
6.7	420	0.3
7.8	420	0.26
8.9	320	0.21
8.9	380	0.28
10	230	0.19
10	350	0.17
11	75	0.08
11	125	0.11
11	275	0.2
11	325	0.19
11	375	0.22
13.3	350	0.16
14.4	150	0.09
16.7	350	0.12
19	150	0.08
19	250	0.095
19	350	0.1
22.2	200	0.075

MAT, mean annual temperature; NSQ, NS Quotient = mean annual rain (mm)/[mean saturation pressure (mm) − mean vapor pressure (mm)]

Here, you are asked to determine the *relative sensitivity of each of these properties to rainfall*. One cannot use the value of the slopes of the equation directly, since there are differing units among the properties. Following Jenny, calculate the various values of the properties at the lowest rainfall (40 cm) and normalize these to 1. Then, calculate the values at 10 cm intervals to 100 cm MAP and convert these to changes relative to the 40 cm values. Plot the data (remember, pH is the negative log of H ion concentration, so you will need to convert to concentrations).

Which property changes the most with rain, and why do you think this might be the case? Which is the most sluggish, and what might be the underlying reason for this?

2 An Overview of the Biogeochemistry of Soils

> ... that the Earth has not always been here – that it came into being at a finite point in the past and that everything here, from the birds and fishes to the loamy soil underfoot, was once part of a star. I found this amazing, and still do.
>
> Timothy Ferris, Seeing in the dark[1]

The chemical and mineralogical composition of soils is most understandable, and interesting, when it is viewed in the context of our cosmic surroundings. The diversity of cosmic chemical processes, and the vast amount of time that has shaped the composition of the universe, has conspired to endow the Earth and its neighboring planets with a relatively unique chemical composition, which in turn has a direct bearing on the composition of soils.

At first glance, it may seem that the Earth's surface is a carbon-rich environment. The global biomass and many products of human invention are carbon-based. Yet, in relation to the solar system in particular, and the universe in general, the Earth is a carbon-, and light element-, depleted planet. The Earth and its soils are dominated by silicates. How this occurred, and its significance to soils, are the subjects of this chapter.

2.1 The Chemistry of the Earth's Crust

The chemical composition of any given soil is the result of a suite of processes that ultimately link that location on the Earth to the history of the universe.[2] The principal processes that have led to the chemical composition of average soil are (1) the chemical evolution of the universe and the solar system, (2) chemical differentiation of the planet, and (3) the effects of soil formation.

The universe is considered to have originated from a "big bang" of an unimaginably dense (10^{96} g cm^{-3}) and hot (10^{32} K) nucleus. About 1 second after the "bang," protons, electrons, and neutrons emerged from the early matter, allowing the formation of H and He. This began the period of element building, in which H and He, and other atomic particles, undergo nuclear reactions to produce elements of greater atomic number and weight. After the first 8 minutes, the mass of the universe is believed to have consisted of approximately 75 percent H and 25 percent He. Today, these two elements still comprise about 99 percent of the mass of the universe. The production of elements of greater atomic number requires a series of nuclear processes that occur as stars undergo long-term changes. Thus, the presence of elements of atomic numbers larger than He are indicative of the fact that

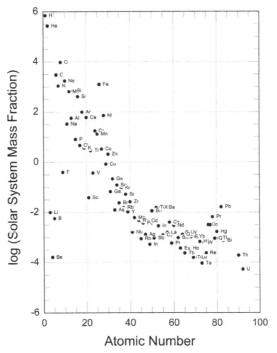

Figure 2.1   The log of the mass fraction (mg kg^{-1}) of elements in the solar system arranged by atomic number. Data from table 1 in E. Anders and N. Grevesse, Abundances of the elements: Meteoric and solar. Geochimica et Cosmochimica Acta, 53: 197–214 (1989).

some of the matter in the universe, and our own solar system, is several generations old, having passed through earlier stars and then having been subsequently distributed throughout space.[3]

The chemical composition of the solar system (which is similar to that of the universe) shows a roughly exponentially decreasing abundance of elements with increasing atomic number (Figure 2.1). Numerous exceptions to this trend exist, but the general pattern of decreasing abundances of progressively heavier elements is related to the series of nuclear reactions needed for more complex elemental synthesis and the time needed for these processes to occur. As illustrated in Figure 2.1, the solar system is dominated by H and He, both of which are nearly two orders of magnitude more abundant than the next most common element, O.

The chemical composition of average crustal rock[4] relative to the composition of the solar system reveals systematic differences, which are related to the events and processes that formed crustal rock from the solar system elemental pool: (1) the accretion of the Earth in relation to its proximity to the Sun, (2) the differentiation of the core, mantle, and crust within the Earth, and (3) late-stage accretion of cometary material. A comparison of the log of the concentration of elements in the Earth's crust with that of the solar system (Figure 2.2) reveals some important patterns that correlate with an element's position on the periodic table (and its resulting chemical characteristics) and the processes that have affected it:

- Earth is depleted in the noble gases (group IIIVA on the periodic table) and H, C, N.
- Earth is enriched in most remaining elements. For these elements, there is a trend for decreasing enrichment with increasing atomic number within a given period, culminating with the highly depleted noble gas of that period.

These trends have been explained in terms of the conditions under which the Earth condensed from a solar cloud and dust. Briefly, the accretion of dust in the early solar system occurred in a disk, which was increasingly hot toward the center (now the location of the present Sun). Elements and compounds of relatively high densities and melting/ volatilization points were preferentially concentrated near the hot center (location of present Mercury, Venus, Earth, and Mars), while volatile elements were concentrated in regions further from the hot interior, an area where the external planets now reside. Brimhall[5] has shown that the apparent periodicity of the relative enrichment within a given row on the periodic table shown in Figure 2.2 can be explained on the basis of the volatilities of the elements: Groups IIIB to IB were concentrated in the refractory early condensate of the Earth (along with Al, Si, and As, which were bound into silicate minerals or metals), while elements in Groups IIB to VIA remained volatilized at relatively low temperatures and were therefore depleted.

In addition to the accretionary processes that fractioned the elemental composition of the crust relative to that of the solar system, there are others that differentiated the Earth into three principal structural components: (1) core, (2) mantle, and (3) crust. Brimhall has shown that core formation concentrated the siderophile elements (Fe, Ni, Co, Ag, W, Mo, Re, Ru, Pd, Os, Rh, Ir, and Au) into the core relative to the mantle, a process that occurred early in Earth's history. The enrichment of these elements in the core is reflected in the fact that these elements are depleted in the crust relative to the solar system (Figure 2.2). During past 3 billion years of Earth's history, the crust has been differentiated from the mantle through igneous processes occurring in the presence of water and CO_2. These processes, which have produced igneous rocks of a generally granitic composition, have shaped the composition of the crust. Brimhall has noted that a comparison of the composition of the crust relative to the primitive mantle shows a progressive enrichment of elements in the crust as one proceeds from Group IV to Group VI on the periodic table (see trend in Figure 2.2 after accounting for the previously discussed depletion of the siderophiles and the volatile elements).

In summary, the Earth's crust, the region of soil formation, is an area dominated by eight elements (O, Si, Al, Fe, Ca, Na, Mg, K). These elements, with the exception of O, are not the dominant elements of the universe or our solar system. This somewhat peculiar starting point for our soils is a function of the Earth's history and the processes that chemically shaped it. We, and our soils, reside in a world dominated by O and Si, products of some past stars and their ultimate cataclysmic demise. It is therefore not surprising that in order to understand the behavior and functioning of soils, one must begin with an understanding of the major compounds formed by these two elements and the remaining six dominant elements. We, and the soils we study, are stardust.

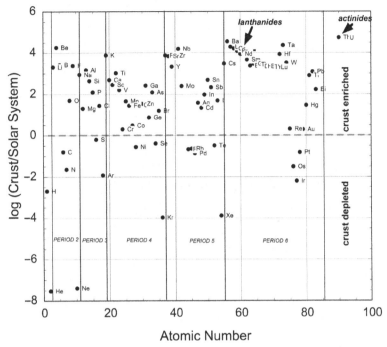

Figure 2.2 The log of the ratio of the average concentration of elements in the Earth's upper crust (mg kg^{-1}) to that of the solar system (mg kg^{-1}). Data on the chemistry of the upper crust from Table 2.15 in S. R. Taylor and S. M. McLennan, The Continental Crust: Its Composition and Evolution, Blackwell Scientific Publications, Oxford (1985) with supplemental data from table 3.3 in H. J. M. Bowen, Environmental Chemistry of the Elements, Academic Press, London, New York (1979). The chemistry of the solar system is taken from E. Anders and N. Grevesse, Abundances of the elements: Meteoric and solar, Geochimica et Cosmochimica Acta, 53: 197–214 (1989).

2.2 Mineralogical Composition of the Earth's Crust

Few, if any, of the important elements of the Earth's crust are found in an elemental state. The elements are combined into compounds, or minerals, which are unique substances that are central to our understanding of soils. The geochemist Claude Allègre wrote:

> Minerals are societies of atoms (elements), rocks are societies of minerals. Society is defined as association by rule, in contrast with mixture, which is random association. The architecture of the crust is clearly divided into levels of organization: atoms, minerals, and rocks.[6]

The rules that dictate the arrangement of elements in minerals are determined by the size of atoms and the way they act on each other by means of their outer electrons. The position of an element on the period table provides insight into how it will tend to gain, lose, or share electrons when combined with another element. Two broad types of chemical bonding have been identified: ionic bonding, in which one element transfers one or more electrons to

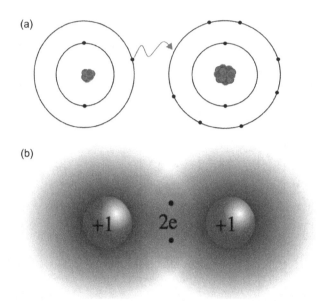

Figure 2.3   (a) Illustration of ionic bonding, where a Li atom donates an electron to a fluorine atom. (b) Two H atoms share electrons in a covalent bond. (Open access from Wikipedia.)

another element, and covalent bonding, in which electrons are shared by two elements (Figure 2.3). Gradations between these two end-members exist as well.

The type of chemical bond that will form between two elements can be discussed in terms of the relative electronegativities of the elements (the power of an atom to attract shared electrons). In general, members of the alkali (e.g. Na) and alkaline earth (e.g. Ca) metals have low electronegativities; the noble gases, halogens, and N and O have high electronegativities; and many of the elements in the middle of the table have intermediate values. If atoms differ greatly in electronegativities (e.g. an alkali metal and a halogen), an ionic bond will form, whereas if there is a small difference in electronegativities, a covalent bond will form. From the perspective of stability at the Earth's surface, covalent bonds between elements are more resistant to alteration than ionic bonds.

The two most abundant elements (in terms of mass) in the crust and soils are O and Si. It is the combination of these two elements that forms the backbone for the mineralogy of the Earth's surface. Silicon and O form a covalent-like bond, although their somewhat moderate difference in electronegativities imparts some ionic character to their bond. This Si-O structure, with one Si atom (which has an ionic charge of +4) surrounded by four O atoms (each with an ionic charge of −2), is called the silica tetrahedron (Figure 2.4a).[7] Aluminum and O also form a bond with both covalent and ionic characteristics. In this case, because of the somewhat larger ionic radius of Al, six O atoms are coordinated around the Al atom, forming an aluminum octahedron (Figure 2.4b). The bonds between the remaining metals (particularly the remaining five most abundant elements on the crust: Fe, Ca, Mg, Na, K) and O in the tetrahedrons are primarily ionic, and in a stable configuration enough of these ionic metal-O bonds are present to impart electrical neutrality to the structure.

(a)

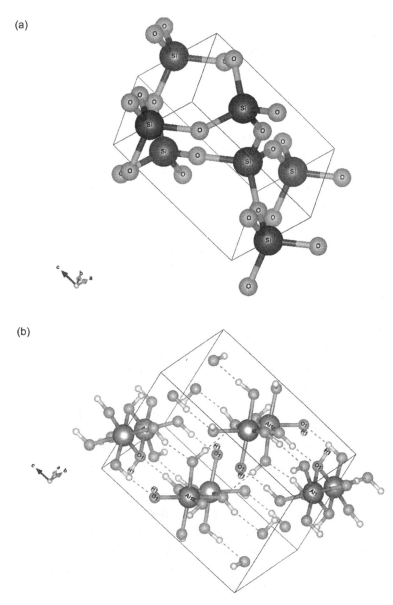

(b)

Figure 2.4 (a) The Si tetrahedron, the backbone of the mineral (shown here) quartz. One Si is bonded to four adjacent O. (b) The Al octahedron, the backbone of the mineral gibbsite (shown). Each Al is bonded with six O. Additional H are required for electrical neutrality of the overall mineral framework. Images created with VESTA: K. Momma and F. Izumi, VESTA 3 for three-dimensional visualization of crystal, volumetric and morphology data, *Journal of Applied Crystallography*, 44: 1272–1276 (2011).

Therefore, it is the O-, Si-, and to a lesser degree Al-dominated chemistry of the Earth's crust that dictates that the basic mineralogical group of the Earth's crust will be the silicates (minerals built of various combinations of silica tetrahedrons). The

Earth's crust is recycled and transformed, over long periods of time, due to intense alteration by heat (which forms igneous rocks) and/or pressure (forming metamorphic rocks). Therefore, to appreciate the mineralogical composition of the crust, an understanding of the array of silicate minerals found in these rocks must be developed.

Silicate minerals are classified on the basis of how "complex" the linkage of tetrahedrons is within the mineral structure. In some silicate minerals, tetrahedrons exist as independent units, bonded together via ionic bonds with interspersed metals. In other silicates, tetrahedrons share O atoms in a three-dimensional network. The type of silicate structure in a given mineral is dependent on its original chemical composition and the conditions (e.g. temperature of a melt) under which it formed. The complete list of linkages, and resulting silicate mineral classification, is:

- *Nesosilicates*: silica tetrahedrons present as independent entities (Si:O = 1:4).
- *Sorosilicates*: two silica tetrahedrons linked together by sharing one oxygen (Si:O = 2:7).
- *Cyclosilicates*: silica tetrahedrons share two oxygens with adjoining tetrahedrons, forming a closed ring structure (Si:O = 1:3).
- *Inosilicates*: tetrahedrons are joined to form either single chains (Si:O = 1:3) or double chains (Si:O = 4:11).
- *Phyllosilicates*: three oxygens of each tetrahedron are shared with adjacent tetrahedrons to form sheets (Si:O = 2:5).
- *Tectosilicates*: every tetrahedron shares its corners with other tetrahedrons, forming a three-dimensional network (Si:O = 1:2).

In terms of soils, and their parent materials, the neso-, ino-, phyllo-, and tectosilicates are the most common.

In general, as the temperature of the mineral formation decreases (from the cooling of a magma), the complexity of silica tetrahedral linkages increases (Table 2.1). At high temperatures, minerals of the nesosilicate group form, in which individual tetrahedrons are linked together by a metal (commonly Mg^{+2} or Fe^{+2}). As the temperature of formation decreases, the increased sharing of O by adjoining tetrahedrons reduces the overall net negativity of the silica matrix, thereby reducing the number of ionically bonded metals needed to establish electrical neutrality. In many of the minerals, Al^{+3} can take the place of Si^{+4} (because of their similar size). This creates a larger net negativity of the tetrahedral structure (−5 vs. −4), increasing the number of ionically bonded metals needed for electrical neutrality.

The structure and composition of the silicate minerals have important implications for how resistant they will be to chemical weathering during soil formation. Some general rules in regard to weathering resistance are:

- **Decreasing Si/O ratios (increased tetrahedral linkages) increase weathering resistance**

Reduced Si/O ratios reduce overall net negativity of the tetrahedral matrix, reducing the need for ionically bonded metals. These metals are very susceptible to removal (i.e. exchange with H^+) by various weak acids in soil solutions.

• *Decreasing Si/Al ratios increase weathering resistance*
The presence of Al^{+3} in place of Si^{+4} creates the need for additional ionically bonded metals, which are susceptible to replacement by acidic solutions.

• *Presence of Fe^{+2} in minerals decreases stability*
The reduced form of Fe is susceptible to oxidation by oxygenated soil solutions. The oxidation alters the size of the Fe ion, physically distorts the mineral structure, and alters the electronic charge, creating the need for the expulsion of a metal.

In light of these simple rules, it would seem that a mineral consisting of no Fe^{+2}, Al^{+3}, and extensively shared tetrahedral O would be the most stable silicate in soil environments, which is illustrated quantitatively by the weathering rate constants provided in Table 2.1. The tectosilicate mineral quartz, which is such a mineral, is commonly unaffected by weathering in many soils and accumulates as a residue over long periods of time. In contrast, olivine, a member of the nesosilicates, is very unstable in most soils.

2.3 Mineralogical Composition of Soils

The mineralogical composition of soils can be quite varied, even as a function of depth in a given soil. However, there are some simple statements that can be made about the distribution of minerals in soils and the types of minerals formed in soils by chemical weathering.

One way to begin to understand soil mineralogy is to consider mineral distribution in relation to mineral size. In many soils, the minerals that dominate the sand (0.05 to <2 mm) and silt (0.002 to <0.05 mm) fractions are minerals inherited from the rock or sediment from which the soil forms (the parent material). If the parent material is derived from igneous or metamorphic rocks, the mineralogy of these size fractions will likely be dominated by primary silicates. The finest size fraction of the soil, the clay fraction (<0.002 mm), is the location where the weathering products of primary mineral commonly accumulate and are concentrated. Mineralogically, these weathering products are dominated by secondary silicates, oxides, and carbonates. The term "secondary" refers to minerals produced by either complete or partial alteration of a primary or secondary mineral, primarily in aqueous solution in the conditions at the Earth's surface (Table 2.2).

2.3.1 Secondary Silicates

Secondary silicates in soils are dominated by members of the phyllosilicate group with smaller amounts of tectosilicate-like minerals. The primary differences between primary and secondary silicates reside in their chemical composition, and as a result, their chemical behavior. The first major means of classifying secondary phyllosilicates is the number of sheets of tetrahedrally and octahedrally coordinated O that the mineral has. A second

Table 2.1 Summary of the structure and physical properties of the primary silicate minerals

Silicate Classification	Tetrahedron Arrangement	Examples	Chemical Formula of Specific Minerals	(+) charge per 100 oxygen	Melting Temperature (C)	Dissolution Rate (log moles $m^{-2}\,s^{-1}$)[a]
Nesosilicates	independent tetrahedra	olivine series	(forsterite)Mg_2SiO_4	100	1890[b]	−7.2
			(fayalite)Fe_2SiO_4	100	1205[b]	
Inosilicates	single chains	pyroxene group	(augite) $Ca(Mg,Fe,Al)(Al,Si)_2O_6$	66[c]	~1200[b]	−11.9
	double chains	amphibole group	(hornblende) $NaCa_2(Mg,Fe,Al)_5$ $(Si,Al)_8O_{22}(OH)_2$	55[b]		−10.3
Phyllosilicates	sheets	mica group	(biotite) $(Mg,Fe)_3$ $(AlSi_3O_{10})(OH)_2$	80	~1100[b]	−12.6
			(muscovite) KAl_2 $(AlSi_3O_{10})(OH)_2$	80	~980[d]	−13.5
Tectosilicates	framework	plagioclase group	(anorthite) $CaAl_2Si_2O_8$	100	1550[b]	−10.3
			(albite) $NaAlSi_3O_8$	50	1100[b]	−11.8
		feldspar group	(orthoclase) $KalSi_3O_8$	50	1150[b]	
		silica group	(quartz) SiO_2	0	867[b]	−13.4

[a] Examples from J. L. Palandri and Y. K. Kharaka, A compilation of rate parameters of water-mineral interaction kinetics for application to geochemical modeling. US Geological Survey Open File Report 2004–1068 (2004). Rates for approximately 25 °C and neutral pHs are given. Note that these are representative values from a large and important data set that should be consulted directly for detailed work.

[b] Data from W. A. Deer, R. A. Howie, and J. Zussman, *An Introduction to the Rock Forming Minerals*, Longman Group, Ltd, London (1966).

[c] Value for endmember with no Al substitution for Si. Value will decrease in proportion to added Al.

[d] From ranges reported in D. S. Fanning, V. A. Keramidas, and M. A. El-Desoky, Micas, chapter 12 in J. B. Dixon and S. B. Weed (eds), *Minerals in Soil Environments*, 2nd Ed, Soil Science Society of America, Madison, WI (1989).

Table 2.2 Table summarizing the structure and physical properties of secondary minerals common in soils.

Mineral Classification	Tetrahedral Sheet Arrangement	Example	Chemical Formula of Specific Minerals	Si/Al + Fe	CEC (cmol(+)/kg) mineral
Phyllosilicates	2(tetra):1(octa)	smectite group	(montmorillonite) $M_x(Al_{3.2}Fe_{0.2}Mg_{0.6})(Si_8)O_{20}(OH)_4$[a]	2	110 (range 47–162)[b]
	2(tetra):1(octa)	vermiculite group	(trioctahedral vermiculite) $M_x(MgFe)_6(Si_{8-x}Al_x)O_{20}(OH)_4$[c]	2	150 (range 144–207)[c]
	1(tetra):1(octa)	kaolin group	(kaolinite) $(Al_4)(Si_4)O_{10}(OH)_8$[d]	1	1 (range 0–1)[d]
Tectosilicates	NA	silica group	(opal) $SiO_2 \cdot nH_2O$)[e]	infinity	0
Oxides	NA	iron oxides	(goethite) $FeOOH$	0	~0 (pH dependent)[f]
			(hematite) Fe_2O_3	0	~0 (pH dependent)
			(ferrihydrite) $Fe_5(O_4H_3)_3$[f]	0	~0 (pH dependent)
		aluminum oxides	(gibbsite) $Al(OH)_3$	0	~0 (pH dependent)[g]
Carbonates	NA	NA	(calcite) $CaCO_3$	NA	~0[h]
Organic Matter	NA	NA	NA	NA	100–900 (pH dependent)[i]

[a] From G. Sposito, *The Chemistry of Soils*, Oxford University Press, New York (1989).
[b] From G. Borchardt, Smectites, chapter 14, in: Dixon and Weed, *Minerals in Soil Environments*.
[c] From L. A. Douglas, Vermiculites, in: Dixon and Weed, *Minerals in Soil Environments*.
[d] From J. B. Dixon, Kaolin and serpentine group minerals, chapter 10, in: Dixon and Weed, *Minerals in Soil Environments*.
[e] Amorphous or paracrystalline.
[f] From U. Schwertman and R. M. Taylor, Iron oxides, chapter 8, in Dixon and Weed, *Minerals in Soil Environments*.
[g] From P. H. Hsu, Aluminum oxides and hydroxides, chapter 7, in: Dixon and Weed, *Minerals in Soil Environments*.
[h] From H. E. Doner and W. C. Lynn, Carbonate, halide, sulfate, and sulfide minerals, chapter 6, in: Dixon and Weed, *Minerals in Soil Environments*.
[i] From J. M. Oades, An introduction to organic matter in soils, chapter 3, in: Dixon and Weed, *Minerals in Soil Environments*.

important property is where charge imbalance in the structure originates. Table 2.2 provides a brief overview of secondary phyllosilicate classification and behavior.

The 2:1 phyllosilicates consist of one sheet of octahedrally coordinated O (and OH) sandwiched between two sheets of tetrahedrally coordinated O. The dominant element in the tetrahedral sheets is Si^{+4}, and varying amounts of Al^{+3} can be found depending on the specific mineral being considered. The octahedral sheet is primarily dominated by Al^{+3}, Fe^{+3}, or Mg^{+2}. The 1:1 phyllosilicates consist of one sheet each of each layer. The second important characteristic of each type of mineral, beyond its sheet arrangement, is the chemical composition of elements around which the O atoms are coordinated. In the case of kaolinite, the most common 1:1 mineral, Si^{+4}, completely fills the tetrahedral sheet, while Al^{+3} similarly fills two out of three positions in the octahedral sheet (the third position remains unfilled to maintain electrical neutrality) (Figure 2.5a). The significance of electrical neutrality is that in solution, kaolinite is unreactive with ions dissolved in the surrounding soil water (with the exception of a small net exchange capability of OH groups at the mineral's edges). In addition, multiple kaolinite platelets (like the other phyllosilicates) can be found stacked on top of each other. In the case of kaolinite, the outer layer of the octahedral sheet (which consists of OH units) is in contact with the outer layer of an adjoining tetrahedral sheet (which consists of O). The H ion is attracted by the two surrounding Os, and as a result the two kaolinite units are joined together by a relatively strong hydrogen bond.

In contrast, the 2:1 minerals all have some points where Al^{+3} substitutes for Si^{+4} in the tetrahedral sheet, or Mg^{+2} (or other metals) substitutes for Al^{+3} in the octahedral sheet. This composition imparts a net negative charge to the mineral structure, which is neutralized by the adsorption of cations from surrounding soil waters. The composition of adsorbed cations takes place in proportion to the composition of ions in the surrounding solution. These adsorbed cations (dominated by Ca, Mg, N, and K – i.e. alkali and alkaline earth metals released from silicate minerals by weathering) are exchangeable and can be displaced by later solutions containing different concentrations and proportions of cations. The net negative charge of 2:1 phyllosilicates is termed cation exchange capacity (CEC), and the alkali and alkaline earth metals that are absorbed are called exchangeable bases (the term "base" coming from the concept that the reaction of these metals with water produces strong bases). The various types of secondary phyllosilicates (and secondary oxides) have characteristic CECs (Table 2.2), which can be used to help determine the general mineralogical composition of the clay fraction from standard chemical data on soils.

The position at which the net negative charge originates in the 2:1 phyllosilicates has an important impact on the mineral's physical behavior. In the smectite group of 2:1 minerals, much of the negative charge originates in the central octahedral layer (Figure 2.5b). As a result, the distance between the charge imbalance and the adsorbed cation is relatively large, and the cation is adsorbed relatively weakly. Because of this weaker attraction to adsorbed cations, the distance between two adjoining smectite mineral platelets can be variable, depending on the valence of the adsorbed cation and its concentration in solution. Therefore, the smectite group is referred to as the expandable 2:1 minerals. This ability for inter-plate distances to vary has important practical implications for soils dominated by smectite minerals. For example, the wetting and drying of these soils causes interlayers to

(a)

(b)

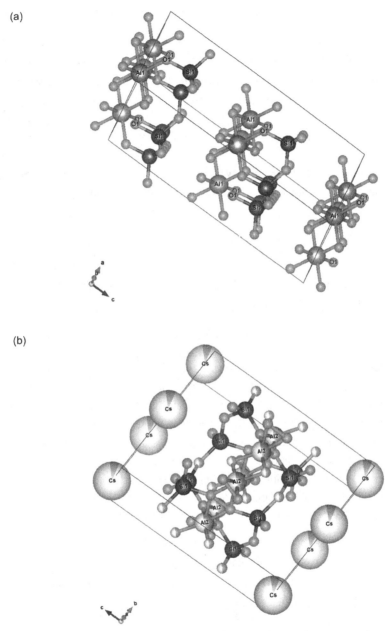

Figure 2.5 (a) An atomic model of kaolinite. The upper layer is the Al octahedral layer, while the lower layer is the Si tetrahedral layer. (b) An atomic model of smectite. The net negative charge (imparted by Al substitution in the outer Si tetrahedral layers and Mg and Fe in the inner Al octahedral layer) is balanced by Cs in the surrounding soil solution. Images created with VESTA: K. Momma and F. Izumi, VESTA 3 for three-dimensional visualization of crystal, volumetric and morphology data, *Journal of Applied Crystallography*, 44: 1272–1276 (2011).

Box 2.1   **Use of Cation Exchange Capacity to Characterize Clay Mineralogy**

The CEC of whole soil samples can be a guide to characterizing general aspects about the mineralogical composition of the soil's clay fraction. Common secondary minerals have a wide range of CECs (Table 2.2) but can be roughly grouped into those with low CECs (approximately 0 to 10 cmol adsorbed cations/kg total soil) vs. those with high CECs (approximately 100–150 cmol(+)/kg total soil). Of course, exceptions to this rule occur (e.g. illite has a CEC of 10–40 cmol(+)/kg). However, using these ranges in CEC, it is possible to use standard soil chemical analyses to determine the relative proportion of high- vs. low-CEC minerals in many soil samples.

Soil CEC data are determined on whole soil samples. It is common in many of these samples that the sand and silt fraction is dominated by primary minerals with little or no CEC. Therefore, as a first approximation, the CEC can be assumed to result from the clay fraction. To convert the CEC of the whole sample to that of the clay fraction, the following expression is used:

$$CEC/kg\ clay = (CEC/kg\ total\ soil)(1/\%clay)(100) \qquad (2.1)$$

To estimate the percentage of the clay fraction dominated by high-CEC minerals, the following mass balance expression may be used:

$$CEC/kg\ clay = X(CEC/kg\ of\ high\ activity\ clay)$$
$$+ (1-X)(CEC/kg\ low\ activity\ clay) \qquad (2.2)$$

where X = the proportion of high-activity (e.g. CEC) clay. This approach to estimating the mineralogical composition of the clay-sized fraction will be compromised by the following (if they apply): (1) the presence of soil organic matter, which has a variable but high CEC, will lead to an overestimation of high-activity minerals, (2) the presence of minerals in the sand or silt fractions with CEC will lead to an overestimation of high-activity minerals, and (3) the presence of intermediate-CEC minerals will add additional uncertainty to calculations. Nonetheless, this approach (Eq. (2.1)) is now used by the United States Department of Agriculture (USDA) to determine the activity classes of soils for the purpose of soil classification (www.nrcs.usda.gov/wps/portal/nrcs/detail/soils/survey/class/taxonomy/?cid=nrcs 142p2_053580 (page now archived)).

also expand and contract, producing noticeable expansion and contraction of the landscape. This has an impact on strategies for the design of housing foundations, road construction, etc. in smectite-dominated soils.

In contrast to the expandable 2:1 minerals, members of the 2:1 nonexpandable minerals (a common example being vermiculite) have most of their charge imbalance originating in the tetrahedral layer due to the substitution of Al^{+3} for Si^{+4}. In these minerals, the distance between the negative charge and the adsorbed cation is relatively small, the adsorbed cations are strongly held, and adjacent vermiculite sheets strongly compete for the inter-layer cations. These adjoining sheets are nonexpandable. Both K^+ and NH_4^+ fit well in the cavity created between Os on the tetrahedral outer layer. Thus, when present in soils, these cations cause the interlayer distance to be minimized, and the two adjoining sheets collapse around the interlayer ions. This then inhibits further exchange with the surrounding soil water.[8]

The final important secondary silicate mineral found in soils is not a phyllosilicate but "opal," a poorly structured member of the silica group of the tectosilicates ($SiO_2 \cdot nH_2O$). In opal, the three-dimensional arrangement of the silica tetrahedra does not extend continuously in space, and as a result, the mineral does not display the degree of crystallinity in normal laboratory identification techniques that quartz, a crystalline silica mineral, displays. There are several types of opal, depending on small differences in chemistry and degree of crystallization. More will be said later on opal formation in soils, but for now it is important to recognize that the Si for opal formation is derived from the weathering of primary silicates in soils and that the Si-containing soil water must be concentrated through evapotranspiration before opal will precipitate from the soil waters. Opal can form in soils (by the mechanism described) or in plant tissue as Si-bearing soil water extracted by the plant is concentrated by transpiration.

2.3.2 Oxides

As discussed later in this chapter, the general chemical trend of soil formation in humid environments is desilication – the dissolution of primary silicates and complete or partial removal of dissolved Si to the streams and oceans. During this process, Al and Fe released from primary minerals tend to combine with water to form one of several possible oxides. In most soils, particularly oxidized and not extremely acid environments, these two elements quickly form oxides and remain relatively immobile (compared with the more mobile Si).

Depending on soil conditions, one of several Fe oxides may form during weathering. A few of these oxides are listed in Table 2.2 and are discussed later, but an important attribute of these minerals in soils is that they impart (even in low concentrations) the variety of ochre, reddish-brown, and red colors characteristic of many soils. In all Fe oxides, a central Fe is surrounded by 6 O or OH groups, forming an octahedron. The packing of these Os can vary, leading to different properties between the various minerals. Three important Fe oxides are:

- *ferrihydrite*: a red, poorly ordered Fe oxide that forms in conditions of rapid Fe release from weathering, low organic matter, and moderate to high pH
- *hematite*: a bright red, ordered Fe oxide formed from altered ferrihydrite in warm, and relatively dry, environments
- *goethite*: a yellowish-brown Fe oxide preferentially formed in acidic, high–organic matter soils

Aluminum released by weathering can be incorporated into secondary phyllosilicates, such as kaolinite, or if Si is vigorously removed from the environment, the released Al will form an oxide, commonly the mineral gibbsite (Table 2.2). In Al oxides, Al is surrounded by 6 O or OH groups, and the packing of these groups determines the specific mineral formed. Gibbsite is concentrated in soils as rainfall and temperature increase (as rates of desilication increase) and is a common mineral in stable tropical soils.

2.3.3 Carbonates

As mentioned earlier, relatively moist and warm environments favor the development of significant quantities of oxides. In contrast, carbonates (primarily the $CaCO_3$ mineral calcite) are favored in environments with low desilication potential – i.e. arid and semiarid environments. Ca released from primary silicates by weathering, or introduced to the soil by $CaCO_3$-bearing dust, moves downward with soil water. As the soil water is removed via evapotranspiration, the solubility product of $CaCO_3$ is exceeded, and calcite will form. Structurally, calcite can be viewed as Ca octahedra (Ca surrounded by 6 O) linked together three-dimensionally by C. In most soils, calcite is tan to white, and its depth in the soil is directly attributable to the water balance of the environment.[9]

2.4 Pathways of Mineral Weathering

To integrate the preceding discussion, a simple diagram of some observed weathering pathways is presented (Figure 2.6). To illustrate these patterns, the United States Geological Survey (USGS) has recently sampled nearly 5000 soil profiles across the USA and conducted mineralogical analyses on the surface and the C horizons. From Figure 2.6, one would predict that as rainfall/temperature (and desilication) increases, the soil surface will shift from weatherable primary minerals to only the most chemically resistant minerals (quartz). Indeed, there is an inverse correlation between Na and Ca plagioclase and quartz in the A horizons of the USA soils, with quartz becoming increasingly rich in the warm and humid southeast portion of the nation (Figure 2.7). Similarly, from Figure 2.5, there should be a general inverse correlation between vermiculite + smectites vs. kaolinite in subsurface horizons where secondary minerals tend to accumulate. Again, the observations correlate with the predications (Figure 2.8). These general trends (mineralogy vs. climate zone) contain additional impacts of other soil-forming factors, particularly landform age, as recently examined by Slessarev et al.,[10] as well as parent material (notice Florida, with an abundance of carbonate-rich rocks).

2.5 Chemistry of Soil Formation

The biological, chemical, and physical processes that combine to form soil are numerous and exceedingly complex. However, starting from a relatively simple perspective, it is possible to begin to understand some important chemical changes that occur to silicate rock or sediment as it is transformed into soil.

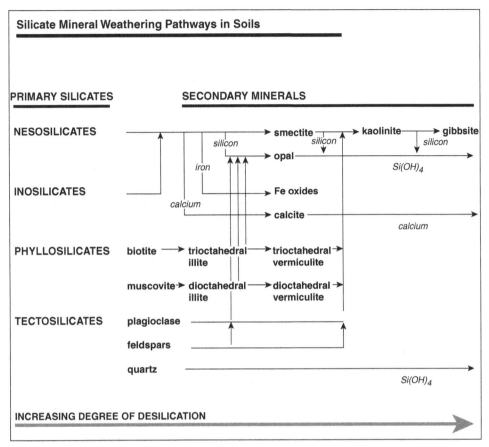

Figure 2.6 A simple diagram of observed weathering pathways in soils. Modified from G. Sposito, *McGraw Hill Yearbook of Science and Technology*, McGraw Hill, New York (1977).

2.5.1 Chemical Weathering

In most regions of the world, liquid water from precipitation passes through the soil zone. This water, particularly when it contains various weak acids (especially carbon dioxide) and complexing agents, is an effective solvent of many of the common silicate minerals. This chemical alteration of the parent material is termed chemical weathering. The variety of specific silicate minerals, and types of chemical alterations, involved under the rubric of chemical weathering is very complex and is a subject that demands a variety of rigorous chemical approaches to be understood thoroughly.[11] However, in regions of the world where precipitation exceeds evapotranspiration, excess water leaves the landscape as streams and rivers. The chemical composition of these water bodies, relative to the landscape being weathered, provides an insight into the chemical changes that soil formation brings to rocks and sediments.

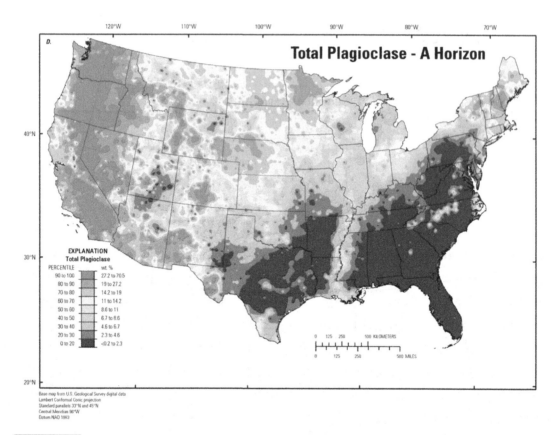

Figure 2.7 The distribution of (a) plagioclase and (b) quartz in A horizons of USA soils. From D.B. Smith et al., Geochemical and mineralogical maps, with interpretations, for soils of the conterminous United States, USGS Open-File Report 2014–1082 (2014).

The chemical composition of natural water is extremely variable, dependent on the lithological composition of the watershed, climate, vegetation, etc. However, estimates of medial values for an array of the elements have been compiled by Bowen and others and serve to illustrate some important aspects of the chemical nature of weathering. Figure 2.9 shows the concentrations of many of the elements in water (as μg of element X g^{-1} of total dissolved solid). These data show that natural waters are dominated by elements of lower atomic numbers, and there is a general decrease in elemental abundance with increasing atomic number. The bulk of the elements are derived from mineral weathering, particularly silicates (see abundance of Si in Figure 2.3).

A useful way to view the chemical data of rivers is to consider it relative to average crustal rocks, the ultimate reactant that supplies the various chemical elements. Here the composition of elements in both freshwater and the crust are compared. When the medial composition of freshwater is plotted relative to average crustal composition, some important points emerge (Figure 2.10). Natural waters are greatly enriched in the elements of the upper portions of the alkali and alkaline earth groups as well as in Si. The bulk of these

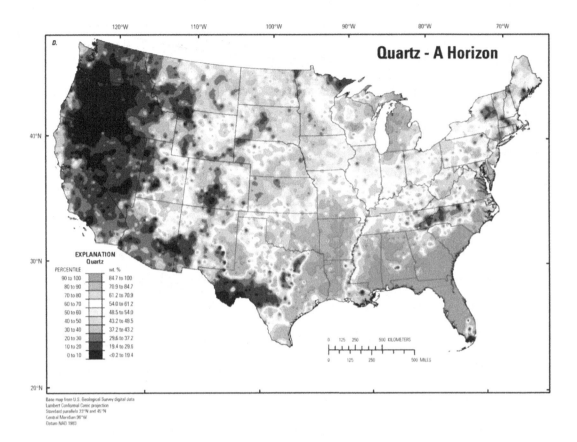

Figure 2.7   (cont.)

elements are derived from mineral weathering, especially the silicates. Additionally, C and N are abundant. Carbon, mainly in the form of HCO_3^- derived ultimately from atmospheric CO_2, is the dominant anion in many natural waters. Nitrogen can be found in several forms, including NO_3^- and dissolved organic N.

In reports or books, the average chemical composition of soil is compared directly with that of the crust. While this exercise reveals general trends, it also produces some unexpected results – such as an apparent increase in the amount of Si in soil relative to the rock and sediment from which it forms. This apparent increase occurs in spite of the fact that Si is one of the major components of river water (Figure 2.11) and that the annual flux of Si to the oceans from chemical weathering is very large (5.6 teramoles Si yr^{-1}).[12] The reason for this apparent discrepancy is that even though Si is lost through the dissolution of many silicate minerals such as olivine, pyroxenes, and plagioclases, some silicates (particularly quartz) are very resistant to weathering and become concentrated in the remaining soil mass as a weathering residue. To more accurately assess chemical changes in soil due to weathering,

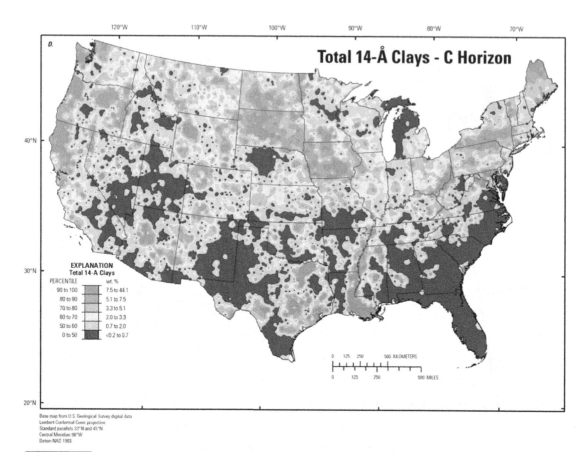

Figure 2.8 The distribution of (a) vermiculite and smectite and (b) kaolinite in C horizons of USA soils. From D. B. Smith et al., Geochemical and mineralogical maps, with interpretations, for soils of the conterminous United States, USGS Open-File Report 2014–1082 (2014).

it is better to evaluate chemical change relative to some immobile index element. Members of the Ti family (group IVB in the periodic table) are to be found in weathering-resistant minerals. In particular, Zr (commonly found in the oxide mineral zirconium) is chosen in many studies as a suitable index element not only because of the resistance of zirconium but also because of the relatively low concentration of Zr in more weatherable minerals. In Figure 2.11, the concentration of elements in soil relative to Zr is plotted relative to the concentration of the same element (relative to Zr) in the Earth's crust. This plot reveals trends consistent with our knowledge of the composition of rivers relative to the crust: (1) Soils are depleted in the alkali metals and alkaline earths (particularly in Mg, Na, Ca, K, Be) and some of the halides (F and Cl). These are the principal cations and anions (with the exception of HCO_3^-) in river waters; (2) soils are depleted in Si, Al, and Fe; and (3) as will be discussed later, soils are enriched in certain elements corresponding to land plants.

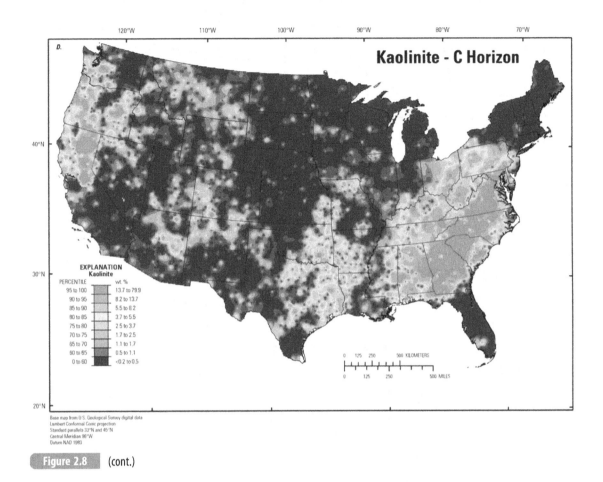

120°W   110°W   100°W   90°W   80°W   70°W

Kaolinite - C Horizon

40°N

30°N

EXPLANATION
Kaolinite

PERCENTILE	wt. %
95 to 100	13.7 to 79.9
90 to 95	8.2 to 13.7
85 to 90	5.5 to 8.2
80 to 85	3.7 to 5.5
75 to 80	2.5 to 3.7
70 to 75	1.7 to 2.5
65 to 70	1.1 to 1.7
60 to 65	0.5 to 1.1
0 to 60	<0.2 to 0.5

0   125   250   500 KILOMETERS

0   125   250   500 MILES

20°N

Base map from U.S. Geological Survey digital data
Lambert Conformal Conic projection
Standard parallels 33°N and 45°N
Central Meridian 96°W
Datum NAD 1983

Figure 2.8   (cont.)

2.5.2 Plant Inputs

Since the emergence of land plants in the Silurian some 400 million years ago, a second process has had a profound impact on soil formation: the addition and cycling of organic matter. Land plants are dominated by C, O, H, and N (Figure 2.12a), with lower quantities of the other elements, which show a general decline in abundance with increasing atomic number. When plants die, their roots are directly incorporated into soil, while portions of their aboveground biomass may enter the soil through biological incorporation or through leaching with water (see Chapter 7). When the concentration of elements in plants is compared with that of the Earth's crust (Figure 2.12b), important concepts emerge. First, plants are five orders of magnitude enriched in C and N relative to the crust. Carbon is, of course, derived directly from the atmosphere during photosynthesis. Nitrogen can be acquired through N_2 fixation or by long-term ecosystem storage of atmospheric N deposition. Secondly, the elements not involved in photosynthesis of major structural components

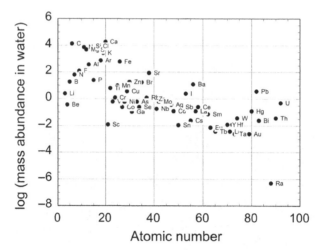

Figure 2.9 The log of the abundance of the elements in natural water.

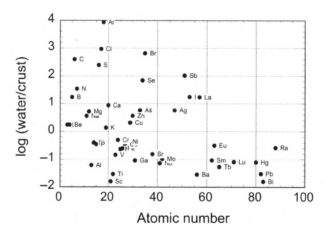

Figure 2.10 The log of the ratio of the chemistry of natural waters to that of the Earth's crust.

are distributed relative to the crust in a remarkably similar pattern to that found in waters relative to the crust (Figure 2.12c). This is understandable because (with some important exceptions) plants "passively" take up dissolved elements with soil water that should, to a first degree, reflect concentrations found in streams.

A general formula of soil formation, based on the gross chemical changes discussed in this chapter, can be viewed as the following general chemical reaction:

rocks + water (+ acids) + plant debris + O_2 = soil + CO_2 + dissolved elements

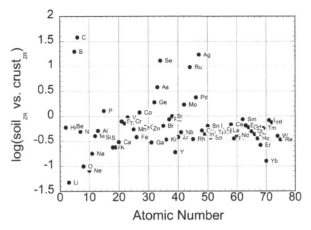

The chemistry of average global soil (normalized to Zr) relative to that of the crust.

Soil parent materials, rocks and sediments, react with water containing various acids (from dissolved CO_2, nitrification, etc.) in the presence of both living plants and their residue. The plants (and the CO_2 and various reactive compounds produced by them), acids, and water are effective agents of chemical alteration. The products of this process include soil (altered rock/sediment plus an array of organic compounds), CO_2 evolved from decomposition of the plant components, and an array of elements dissolved in the soil water. Most importantly, the preceding general equation illustrates that soil formation is a process that does not occur in isolation but is intimately linked to the chemistry of the Earth's surface: the CO_2 content of the atmosphere and the chemistry of streams and oceans.

In subsequent chapters, more rigorous methods of quantifying the rates, processes, and effects of weathering on both soils and the Earth's chemistry will be introduced, as well as the mechanisms by which plant inputs are added to soils and eventually transferred back to the atmosphere as CO_2. These more detailed chapters will also reveal that one of the important characteristics of soils is that they commonly have pronounced variations in properties with depth, changes that are visibly detectable macroscopically in the field. The focus of the next chapter is to recognize how these vertical variations are expressed, and how to recognize them, for they comprise one of the most important attributes of soils.

2.6 Summary

The Earth, and its soils, are stardust, whose unique chemical and mineralogical composition is explainable in light of processes that have occurred over billions of years. The Si and O backbone of the Earth, and its soils, is just the beginning of interesting processes that begin to occur once water and life are added to the system. Based on the location of elements in

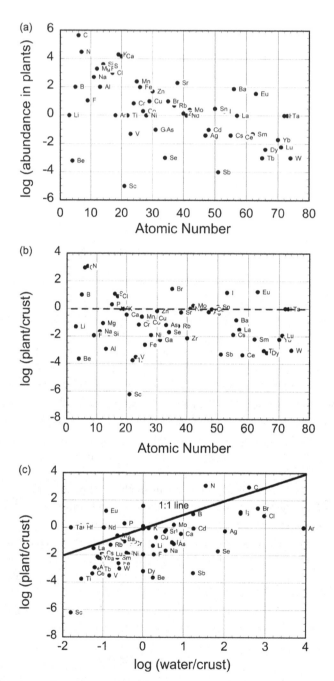

Figure 2.12 (a) log of the elemental abundance of natural waters vs. atomic number, (b) the log of the Zr normalized ratio of plants to average crust, and (c) the relationship between plant/crust and water/crust (log normalized) elemental compositions.

the periodic table (discussed more in Chapter 8), one can understand and predict the mobility, and cycling, of elements through the planet's chemical cycles.

Soil, as we commonly know it, results from small to large alterations of the original parent material by chemical and biological processes. Yet, as discussed in later chapters, the magnitude of both of these processes, driven by the factors of climate and potential biota, are enormously variable on Earth, and in the case of life, may in some locations reach levels that approach zero. Whereas this chapter introduces the reader to the "chemistry of soil biogeochemistry," the next chapter provides insight into its biology and the unique ways that biochemistry has altered our planet and its soils.

2.7 Further Reading

An essential document for biogeochemists is L. Bruce Railsback's paper An earth scientist's periodic table of the elements and their ions, *Geology*, 31: 737–740 (2003). A corresponding website provides additional material that expands on this paper. The paper helps students and professionals alike look at the periodic table of elements anew, with a perspective that helps one understand the behavior of elements and their ions in weathering environments. This paper will be examined again in Chapter 8 on Chemical Weathering, but the reader is encouraged at this point to read the paper and ponder over the expansive and thought-provoking table.

2.8 Activities

2.8.1 Human Chemistry vs. the Stars and Planets

It is commonly stated that we (humans) are stardust. But it's also important to consider that we evolved in an aqueous environment on a rocky planet, and thus, we may differ in many ways from the starting assemblage of cosmic elements. Additionally, some elements on Earth induce physiological toxicity.

We have an enormous and valuable amount of data at our disposal to explore chemical questions. In the accompanying Excel spreadsheet are the "average" composition of humans, water, crust, and the solar system. The human data is from Wikipedia, and the water and solar system are from data used in the textbook.

1. Plot the log (ppm) of humans vs. the log (ppm) of the solar system. What is the apparent relationship, and is this what you might have anticipated?
2a. Plot the $\log_{10}$(human/solar system) mass fraction (in ppm) vs. atomic number. On plots make sure to label zones of enrichment vs. depletion of humans vs. solar system.

2b. For elements #1–20, make a table of relative enrichment/depletion (in humans) relative to the solar system, and (from web search) discuss whether (and why) the element is biologically important or toxic (or has no known use).

3. Calculate the ionic potential (NOT ionization potential) (or find it on the web) for the common ionic form of the first 20 elements. (Railsback provides ionic radius and charge in his periodic table.)

4. Calculate $\log_{10}$(water/crust) for these 20 elements and plot this vs. ionic potential. Discuss the patterns you observe vs. the predictions made by Railsback, and discuss possible reasons for any discrepancies.[13]

5. Plot the log(water mass fraction) vs. log(human mass fraction). Discuss and provide a brief interpretation of what you observe.

2.8.2 Explore Soil Minerals

The examples of mineral structures in this chapter are just introductory examples. In order to explore the diversity of soil minerals, and their chemistry and structure, the Virtual Museum of Minerals and Molecules (P. Barak and E. A. Nater, The Virtual Museum of Minerals and Molecules: Molecular visualization in a virtual hands-on museum, *Journal of Natural Resources and Life Sciences Education*, 34: 67–71 (2005)) is a highly recommended tool: https://virtual-museum.soils.wisc.edu.

Examine examples from the major primary mineral groups, and note where the ionic bonds are located and how their susceptibility to attacks by acid changes with mineral structure. Use these illustrations to fully understand the differences between the secondary phyllosilicates.

3 | The Biology in Soil Biogeochemistry

3.1 Microbiology and Soil Processes

One of the most exciting aspects of modern Earth science is the growing focus on the way that biological communities of plants and microbes, and their metabolic processes, have profound impacts on soils and Earth surface processes – and *vice versa*. It is an area of rapid discoveries and advancements. The interplay and feedback between life and the physical planet is a complex system, one that results in what are called *emergent properties*, properties or processes unexpected from the makeup of the system. In other words, the system is more than the sum of its parts. This global complex system was termed the *Biosphere* by the Russian chemist and earth scientist Vladimir Vernadsky in his 1926 book.[1] Soils, a segment of the Biosphere, share many of the complexities and emergent properties of the global Biosphere of which they are a part.

As an introduction to the biology of soil biogeochemistry, this chapter will review our most recent information on the tree of life and the molecular processes that it controls. This chapter examines how these metabolic processes emerged and changed over time and how today they are distributed across the planet and contribute to the habitability of Earth. The focus is largely on the microbial components of life, in that they drive many of the biogeochemical cycles considered.

An inescapable aspect of this field is that it is changing at an incredibly rapid pace as the cost of genetic analyses decreases, the technology improves, and more scientists become engaged in this vibrant scientific field. The research that forms the backbone of this chapter is all relatively recent, and a decade or so ago, much of this was unknown. In another decade, some of this will likely seem antiquated. However, it is nonetheless instructive to wade into these rapidly moving intellectual waters and get a sense of the present lay of the intellectual landscape.

3.2 The Diversity of Life on Earth

The diversity of life on Earth is staggering, and it's important to recognize that major changes in our understanding of life and its breadth have occurred in only the last few decades. It is now agreed that there are three domains (fundamental categories) of life on Earth: *Bacteria, Archaea*, and *Eukaryotes* (Figure 3.1).[2] Bacteria and Archaea are prokaryotes, unicellular organisms that lack a membrane-bound nucleus and other biochemical machinery. Eukaryotes have a nucleus and can be, and are most commonly, multicellular.

The "peacock tail" of taxonomic "feathers" in Figure 3.1 reveals many important points. First, most of the feathers are in the Bacterial domain (92 phyla), which highlights that life on Earth is dominated, species-wise, by bacteria. Second, the Eukaryotes, multicelled life that includes us, is located close to the single-celled domain of Archaea on the tail. This is because it is now largely believed that Eukaryotes may have emerged when an ancient Archaeon engulfed a Bacterium, and both survived and merged, resulting in multicellular life, a process called endosymbiosis.[3] Third, and most importantly for soil biogeochemistry, feathers signified by a dot in the figure have never been seen in living form. In other words, science only knows of them through their DNA. Most of the organisms in soil are at present unculturable in the lab, and scientists never knew they existed until the recent revolution in molecular biology. There is much to discover about what they all do, how fast they do it, and ultimately what their fundamental roles are.

Microorganisms make up an important fraction of the total organic C in soils. The mass of the total soil organic matter that is comprised of microbial biomass is commonly determined in the lab by fumigation of a soil sample (to lyse the microbial cells) and an extraction step.[4] A recent compilation of published biomass estimates shows that on average, microbial biomass is about 1.2 percent of the total C, but that greater fractions of N (2.6 percent) and P (8.0 percent) are contained in living microbes, revealing their importance in biogeochemical processing and cycling.[5] While the percentage of total soil C in living microbes may seem small, globally it totals roughly 37 Gt C, equivalent to nearly 4 years of anthropogenic emissions to the atmosphere. It is these organisms, and their genes, that mediate the vast global cycles of elements that pass through soils and into the atmosphere and the hydrosphere.

3.3 The Biological Imprint on Earth's Biogeochemistry

It has been stated[6] that the chemistry of abiotic geochemical reactions tends to be dominated by acid/base reactions (the transfer of protons, H^+). For example, the weathering of silicate minerals is mechanistically driven by proton attack and exchange with cations at the mineral surface, and expressions to describe the weathering rates of silicate minerals commonly incorporate proton concentrations (e.g. pH). In contrast, reactions that include biology are dominated by redox reactions, which involve transfers of both protons and

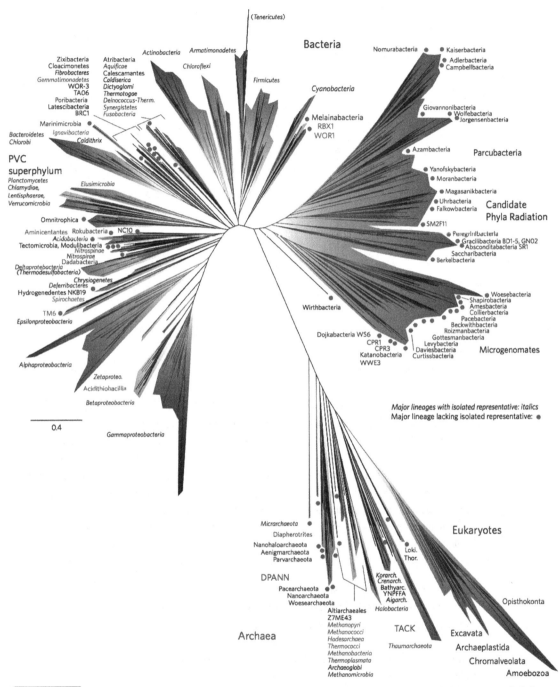

Figure 3.1   The present view of the tree of life represented by sequenced genomes. The tree includes 92 Bacterial phyla, 26 Archaeal phyla, and all 5 of the Eukaryotic supergroups. The major lineages are assigned arbitrary colors and named with well-characterized lineage names in italics. Lineages that have not been isolated are identified with non-italicized names and dots. The Candidate Phyla Radiation are assigned one color because they are composed of organisms that have never been isolated, and they are still being defined at lower taxonomic levels. From L. A. Hug et al., A new view of the tree of life, *Nature Microbiology*, 1: 1–6 (2016).

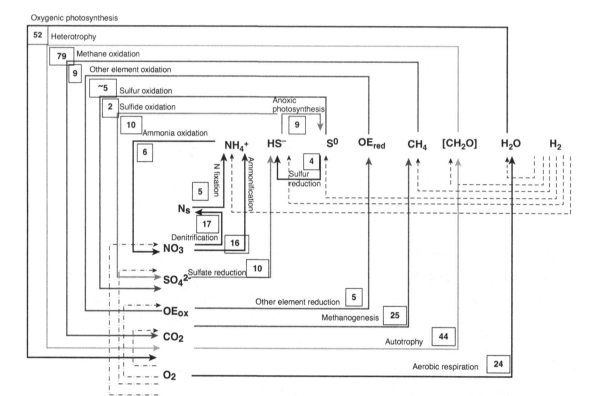

Figure 3.2 Diagram showing the major electron pathways on Earth and in soils. The number of genes involved in each step are given in the boxes. The upper left pathways are oxidative while the lower right are reductive. Solid lines are redox couples, and dashed lines represent participation of O_2 or H_2. Abbreviations are: A = ammonification; AP = anoxygenic photosynthesis; AR = aerobic respiration; AU = autotrophy; D = denitrification; Ex_{ox} = other elements oxidation; Ex_{red} = other elements reduction; H = heterotrophy; M = methanogenesis; MO = methane oxidation/methanotrophy; N/AO = nitrification/ammonia oxidation; NF = nitrogen fixation; OP = oxygenic photosynthesis; SDO = sulfide oxidation; SO = sulfur oxidation; SR = sulfur reduction; STR = sulfate reduction. From B. I. Jelen et al., The role of microbial electron transfer in the coevolution of the biosphere and geosphere, *Annual Reviews in Microbiology*, 70: 45–62 (2016).

electrons from what are a small set of chemical elements of enormous biogeochemical importance to soils: C, N, O, H, S, and (to a minor degree) P (Figure 3.2). Life evolved over geological time by creating genes that encode the synthesis of proteins that mediate and catalyze redox half-cells, which underlie the energy-transducing metabolic pathways. The forward and reverse steps in elemental oxidation/reduction form and maintain elemental cycles.

When the various important metabolic pathways first emerged in Earth's history is difficult to accurately establish due to the complexities of the geological record and due to horizontal gene transfer between organisms. This has allowed biological sharing of evolutionary innovation and made the microbial world a pool that serves to preserve life's

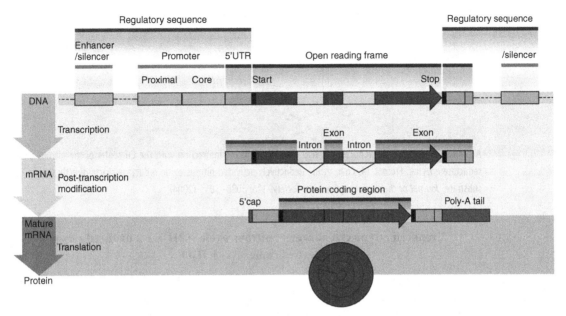

Figure 3.3 The structure of a prokaryotic operon of protein-coding genes. The sequence of DNA segments controls when expression occurs for the protein-coding regions (red). The yellow regions regulate the transcription of the gene into mRNA, and in the mRNA, the blue untranslated regions regulate the translation into the final protein products (red circles). From T. Shafee and R. Lowe, Eukaryotic and prokaryotic gene structure, WikiJournal of Medicine, 4: doi:10.15347/wjm/2017.002 (2017).

metabolic innovation through time and through multiple global cataclysmic events. One important feature of the biosphere is that life has changed the chemistry of the Earth's surface, and this changed environment has in turn changed life itself.

The way in which microbes ultimately influence soil biogeochemistry (respiration, N and P cycling, etc.) is through genes, sections of the microbial DNA or RNA that code for a molecule, such as an enzyme, that has a specific biochemical function. Complex chemistry thus underlies the basis and functioning of life. An illustration of how an organism, in this case a prokaryote, creates proteins that facilitate metabolic processes is shown in Figure 3.3. A segment of the DNA sequence is illustrated in the upper row of arrows. Read from left to right, the sequence initiates, codes, and completes the transcription (reading and replication as messenger RNA, mRNA) of the code for a specific protein. The mature mRNA serves as a template for the synthesis of a new protein by ribosomes, which consist of RNA and proteins that add amino acids to the growing peptide chain. Genes can be regulated by various environmental or physiological factors and are expressed when needed. Proteins are macromolecules of amino acids that serve a large number of functions. Many contain transition metals, capable of oxidation and reduction (e.g. the transfer of electrons), which help facilitate and catalyze a metabolic reaction. An example is nitrous oxide reductase, encoded by the *nosZ* gene, which contains Cu at its center (Figure 3.4):

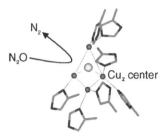

Figure 3.4 A simple schematic of the reduction of N_2O to N_2 through the interaction with the Cu center of the nitrous oxide reductase enzyme. From J. M. Chan et al., Reductively activated nitrous oxide reductase reacts directly with substrate, *Journal of the American Chemical Society*, 126: 3030–3031 (2004).

2 reduced *c*-type cytochrome + nitrous oxide + 2H$^+$ → 2 oxidized *c*-type cytochrome + N$_2$ + H$_2$O

The enzymes that are involved in electron transfers, which drive the cycling of C, N, O, and S, are called oxidoreductases. Despite the diversity of life, the entire network of electron transfer reactions on Earth involves fewer than 400 gene families, and these genes have been shared across the domains of life through vertical and horizontal gene transfers.[7] Oxidation/reduction cycles are accomplished by the combination of two half-cells, oxidizing/reducing couples. The measure of the tendency of a chemical species to gain or lose electrons is called its redox potential, the potential to gain or lose electrons relative to a standard hydrogen electrode. Standard reduction potential (E_o) in nature ranges from ∼−750 mV (highly reducing) to >1000 mV (highly oxidizing) (Figure 3.5). Enzymes are produced by microbes with their midpoint potentials developed to efficiently catalyze the reactions of interest (Figure 3.5).

Over geologic time, life oxidized the Earth's surface. This was accomplished by the evolution of photosynthesis, which uses water, an electron donor, to create carbohydrates and O_2. This is a global redox half-cell. The counter process is respiration, which reduces the O_2 back to water. The evolution of photosynthesis created significant quantities of O_2 and carbohydrates. Over geological time, some of the organic C was buried in sediments and rock, removing part of the global half-cell reductant. As a result, the Earth's surface oxidation state became more oxidized, and the nature of the abundant chemical species for redox reactions subsequently also changed to more oxidized redox pairs (Figure 3.5).

In turn, life has had to adjust to the increasingly oxidized atmosphere, soils, and oceans and has been forced to evolve new metabolic pathways to take advantage of, or simply survive, these changed conditions.[8] One key change is that Fe^{+2}, which was abundant prior to the oxygenation of the atmosphere by photosynthesis and organic C burial, was a common cofactor in many enzymes. Following the oceanic oxygenation, Fe became far less abundant in oxygenated waters (converted to Fe^{+3} and one of the many Fe oxides (Chapter 2)), and many microbes appear to have then begun to utilize Cu, Mo, and other metals, which were both more relatively abundant and also had the advantage of creating enzymes with higher midpoint oxidation potentials. For example, the nitrous oxide reductase enzyme discussed earlier mediates a reaction at a very high E_o and utilizes Cu in its catalytic center.

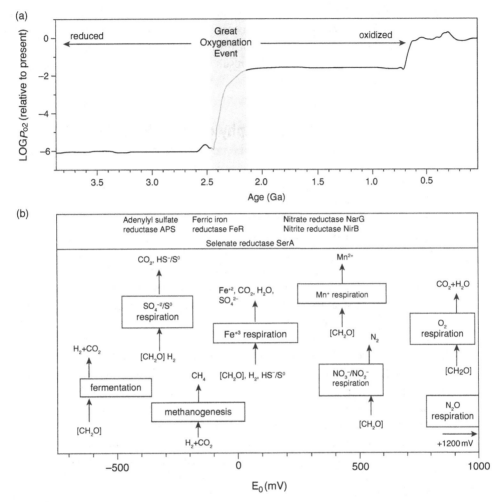

Figure 3.5 (a) The changes in the Earth's oxygen content and the major electron donors and acceptors over Earth's history (relative size of font illustrates relative abundance). (b) Standard redox potential of biologically important redox pairs. (b). From B. I. Jelen et al., The role of microbial electron transfer in the coevolution of the biosphere and geosphere. *Annual Reviews in Microbiology*, 70: 45–62 (2016).

While the oxidation/reduction cycles of the elements of soil biogeochemical interest (Figure 3.2) are commonly considered individually, they are all complexly linked in a number of ways. First, the oxidation of one cycle (say S) may be linked to the reduction of, for example, NO_3. Second, in the case of N, for example, a number of different enzymes are involved in the complete metabolic cycle, not all of which are commonly found in any one organism, and thus, complex interactions of organisms and their enzymes are required to fully complete the cycle. The sweep and impact of biology on Earth is not restricted to elemental cycling and behavior, and it ultimately impacts the mineralogical makeup of the planet. As discussed in Chapter 2, with its focus on the silicates, mineralogy might seem to be largely an inorganically mediated process. However, due to redox processes, elemental concentration by biology, and other biotically mediated processes, it has been estimated that >4000 mineral species on Earth may result directly or indirectly from biogeochemical processes.[9]

3.4 The Geography of Soil Microbiology

The soil environment that the microbes live in is defined by the assemblage of state factors for that location. In Chapter 1 the state factor model was introduced. Here, the biotic factor is considered in more depth. Following the publication of "Factors of Soil Formation" in 1941, Jenny continued to struggle deeply with the biotic factor and what it means. After nearly two decades, he elaborated a much deeper definition of the biotic factor.[10] Briefly, Jenny recognized that at any site (e.g. ecosystem), the plants and microbes that are alive are commonly just a subset of the genetic flux into that system, those that can survive and thrive within the array of state factors that control the site. The life that thus lives in a given soil ultimately has direct impacts on the properties and functioning of the soil, and the soil environment in turn impacts the makeup of the community of life. Thus, like the Earth's biosphere as a whole, soil and its life coevolve with each other. The truly *independent* biotic factor is the flux of seeds, spores, and microbes from the surrounding landscape, which can reflect inputs from many kilometers or more away. Jenny called this the *biotic potential*, which may or may not be fully expressed due to soil conditions. Thus, what lives at a site may be a small subset of what initially enters the system.

There are some interesting points or implications of this concept. First, it does not directly support a hypothesis in microbiology that "everything is everywhere, (but) the environment selects."[11] Jenny recognized that the environment selects what can actually live in a system, but the unique conditions within the system are also evolutionary controls and can facilitate, over appropriate time intervals, considerable evolutionary change and innovation at a local scale. This is likely being empirically verified by the ongoing discoveries of what appear to be endemic microbial species in unique environments.[12] Second, to understand the controls of soil properties on biota, one must correctly distinguish between dependent and independent variables. For example, an easy-to-measure soil property is pH, and in a number of analyses, pH has been found to be strongly correlated with some microbial populations. However, soil pH is not, at least in most cases, an independent variable. It is a soil property that changes over time[13] in response to the other state factors, particularly the amount of rainfall available for leaching. This was recently articulated by Slessarve et al.,[14] who showed from 20,000 soil measurements derived from databases that the pH increased as the mean annual precipitation (MAP) minus the potential evapotranspiration (PET) declined. Soil pH, like the microbial community itself, is a dependent variable that responds to the other, truly independent, state factors, in this case climate. Thus, more correctly, both soil pH and microbial populations are codependent (to a first order) on climate. The concept that soil microbiological communities are dependent variables is important for developing a better predictive capability of how communities vary with state factors[15] and the resulting changes in metabolic functioning across natural landscapes, and how they might change as a result of changes in climate or land use. The following is a brief view of our present understanding of the geography of soil microbiology.

As illustrated in Figure 3.1, bacteria are the largest of the three domains of life. During the past decade, targeted soil sampling and the use of 16s rRNA analysis to develop microbial phylogenies has resulted in the first maps of the spatial variation in soil bacteria.[15] This study identified 511 major phylotypes that make up 41 percent of the total abundance in soils. However, over 24,000 phylotypes comprise the remaining bacterial abundance. More simply, the dominant phylotypes (2 percent of the total) were dominant across continents (Figure 3.6). The authors found that about half of the 500 major phylotypes could be predicted from environmental variables (e.g. pH, aridity indices, etc.) and classified six environmental clusters (Figure 3.7e), whose distribution across the globe is projected in Figure 3.7a–d.

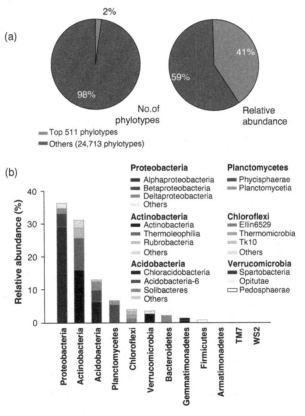

Figure 3.6 (a) Percentage of bacterial phylotypes and relative abundance for soils around the world. (b) The taxonomic composition of the dominant phylotypes of bacteria. From M. Delgado-Baquerizo et al., A global atlas of the dominant bacteria found in soil, *Science*, 359: 320–325 (2018).

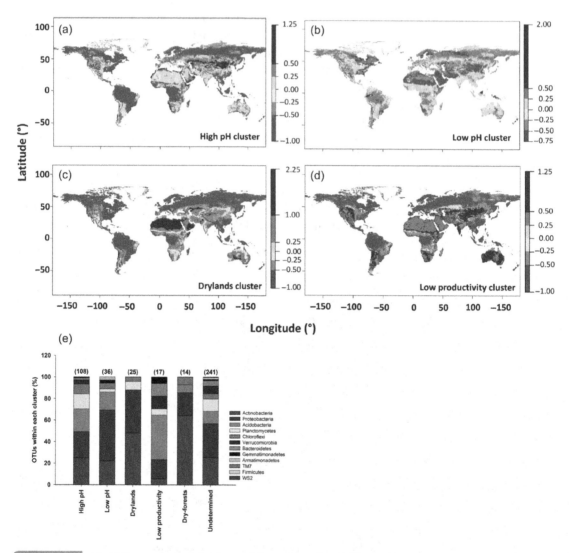

Figure 3.7 (a)–(d) The standardized abundances of four major ecological clusters of bacteria on a geographical basis. (e) The taxonomic makeup of each cluster. From M. Delgado-Baquerizo et al., A global atlas of the dominant bacteria found in soil, *Science*, 359: 320–325 (2018).

The remarkable ability to map and predict a first-order bacterial biogeography is likely to greatly impact developments in soil biogeochemical models. First, as metabolic functions of the phyla become better understood through metagenomic investigations, it should be possible to improve models of C and N cycling, for example, by explicitly improving the representation of enzymatic or metabolic parameters. One example of this is a study of the comparison of functional genes found in desert soils (warm and cold) vs. nondesert soils.[16] Using metagenomic analyses, the authors identified the most important functional gene differences between deserts and nondeserts and categorized them into 15 key

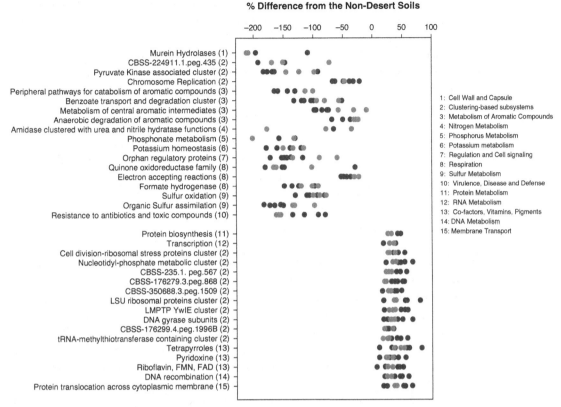

Figure 3.8 Specific gene categories that are relatively more or less abundant in desert soils than in nondeserts, showing only genes that are significantly different. From N. Fierer et al., Cross-biome metagenomic analyses of soil microbial communities and their functional attributes, *Proceedings of the National Academy of Sciences*, 109: 21390–21395 (2012).

biogeochemical categories (Figure 3.8). Deserts are lower in genes associated with N, K, and S cycling, presumably due to the aridity and slow nutrient cycling rates. Desert soils also have lower antibiotic capabilities, which were hypothesized to result from lower microbial competition in these relatively low–microbial biomass environments.

It should be noted that soil properties depend also on time. Recently, 16 soil chronosequences were studied using marker gene amplicon methods to characterize how the diversity of the three domains of life varies with soil age.[8] Since soil changes over time are also impacted by the climate, the chronosequences were grouped into high- and low-productivity (wet/hot vs. cold/dry) clusters. In the cold/dry environments, soil microbial diversity increased over time, mirroring increases in plant cover. In the warm/wet chronosequences, soils tended to become acidified (and more nutrient poor) over long periods of time, and as a result, soil biodiversity declined. These changes over time reemphasize the fact that soil properties and microbial communities are both dependent variables and that they tend to coevolve with each other over time and space.

The groups and functioning of natural soil microbial populations can dramatically change with human intervention. Cultivation, as discussed elsewhere in this book, imparts a profound change to the soil system: removal of native flora and fauna, radical physical disturbance of soil structure, and addition of nutrients and other compounds. Probably few ecosystems on Earth have been as heavily altered by cultivation and human intervention as the former tall grass prairie of the central USA.[17] Today, only small, scattered remnants of unplowed prairie remain within this otherwise agricultural landscape. Fierer and others[18] measured the diversity and the functions of microorganisms in remaining undisturbed soils in the prairie ecosystem. They discovered that the natural soils are dominated by *Verrucomicrobia,* a bacterial phylum. Metagenomic analyses reveal that this phylum is characterized by relatively abundant carbohydrate-metabolizing genes and less abundant N cycling–associated genes. The authors proposed that this phylum is adapted to the some-what recalcitrant nature of prairie flora, which requires more C cycling than N cycling capabilities for nutrient access. A related research paper[19] demonstrated that adding N fertilizer quickly reduces the abundance of *Verrucomicrobia* and other species and reduces the rate of soil respiration.

Verrucomicrobia is a relatively poorly studied phylum of bacteria, and this phylum, as well as the array of microbes in undisturbed soils, likely contains metabolic features of considerable human interest, which we may be altering or losing through our activities. The human cultiva-tion and use of soils is the "below-ground" equivalent of the disruption of tropical rainforests in terms of impacts on biodiversity. Why is this important? For example, due to growing antibiotic resistance from the excessive use of existing antibiotics, untreatable human infections are increasing and may increase 10-fold by 2050. A number of antibiotics have been derived from cultured soil microorganisms over the past century, but over time the rate of discovery of novel natural products has declined as the relatively small number of culturable organisms have been exploited.[20] Using genome-resolved soil metagenomics (the analysis of the entire DNA of a given soil), a method that has only recently become possible but provides a complete perspective of the entire microbial structure, Crits-Christoph et al.[21] found that the major bacterial phyla in soils from California (Figure 3.9) all contained a large number of previously unrecognized gene clusters that code for non-ribosomal peptide synthetases (NRPSs) and polyketide synthases (PKSs), enzymes whose products include many antibiotics, antifungals, and immunosuppressants. A high percentage of these genes were found to be active upon the addition of water and a carbohydrate, revealing unrecognized potential diversity of natural products for potential medical applications. The authors noted or speculated that the number of such gene clusters was related to the degree of interaction these organisms experience and is thus linked to ecological competition. How this genetic reservoir changes with human disturbance of soils is presently not well studied, but research on the Great Plains soils suggests that the changes are significant for many processes, some of which we are only beginning to understand.

3.5 Parameterizing Biology in Biogeochemical Models

As metagenomic studies evolve to quantify the genes that control the various biogeochem-ical cycles, there will be a renewed effort to parameterize the metabolic capacities of soils more clearly and explicitly in biogeochemical models.

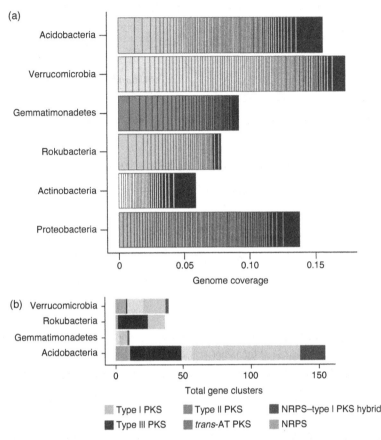

Figure 3.9   (a) Mean relative abundances of the genomes of 60 soil samples as determined by sequencing coverage of the genomes. (b) NRPS and PKS gene clusters from each phylum. From A. Crits-Christoph et al., Novel soil bacteria possess diverse genes for secondary metabolite biosynthesis, *Nature*, 558: 440–444 (2018).

In this book, the approach at this relatively introductory level is to use generic constants that implicitly include whatever the biological controls and impacts are. An example, which is discussed at greater length later, is the soil C cycle:

$$\frac{dC}{dt} = I - kC \tag{3.1}$$

Where C = total soil C (mass/vol), I = C input rates (mass/area*time), and k = first order decay constant of C (time^{-1}).

This simple expression implies that soil C is decomposed to CO_2 at a rate proportional to the total pool size. It also implies that microbes, in a given location, respond linearly to changes in C. If studies are performed at a variety of locations, the climate dependence of k can be determined. This approach, which is embedded into a number of important earth system models, implicitly lumps the microorganisms, which biochemically mediate the

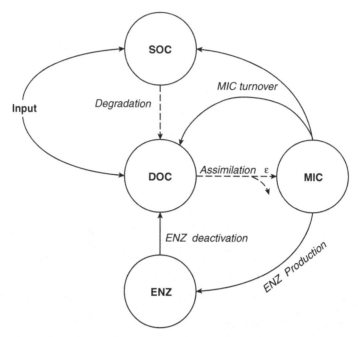

Figure 3.10   A microbial-explicit soil C cycling model. SOC = soil organic C; DOC = dissolved organic C; ENZ = enzymes; MIC = microbes. Degradation of SOC to DOC is governed by ENZ and temperatures and assimilation by the size of the MIC pool. From W. R. Wieder et al., Explicitly representing soil microbial processes in Earth system models, *Global Biogeochemical Cycles*, 29: 1782–1800 (2015).

decomposition process, into a "black box" of the decomposition constant. To begin to consider the microorganisms more mechanistically, one might begin by portraying the genetic information (depending on the species) that codes for the production of enzymes that are involved in specific or broad arrays of organic matter breakdown. However, these genes may not immediately or directly impact biogeochemistry. Some genes may be suppressed, and even if the genes are active, the resulting enzymes are subject to diffusion rate limitations in soils before they can interact with the target compounds. Additionally, they respond to both temperature and moisture changes. A diagram of a soil C model explicitly including microorganisms is shown in Figure 3.10.

A nonlinear and more explicit representation of biological processes in C cycling might be[22]

$$\frac{dC}{dt} = I - C \cdot \left(f(C, E) \right) \tag{3.2}$$

where some function (f) describes the nonlinear response of enzyme availability (E) and microbes to C amounts. Temperature dependence, diffusion rate parameters, etc. could also be incorporated into these expressions. While the inclusion of more parameters can in turn make models more representative of observations, it is presently difficult to realistically determine the values of the parameters in most locations. For example, enzyme pool size is determined by extractive methods and at this stage not well linked to genetically controlled

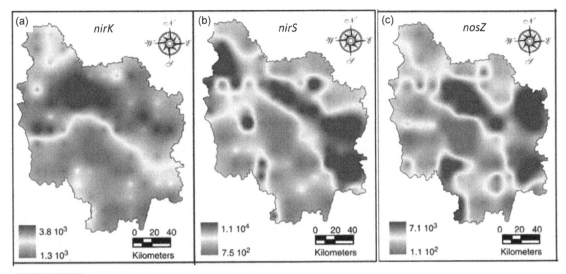

Figure 3.11 Maps of the abundances of N cycling genes in soils from Burgundy, France. The shading scale indicates gene copy number per nanogram of DNA. From D. Bru et al., Determinants of the distribution of nitrogen-cycling microbial communities at the landscape scale, *The ISME Journal*, 5: 532–542 (2011).

capabilities. It is likely that a number of important processes in soil, for example nitrous oxide (N_2O) production and consumption, are affected by the waxing or waning of differing microbial populations that may possess one, or several, key enzymes in the overall process. Accurately determining these enzymatic ratios involves various metagenomic approaches, and the changes in these ratios likely change over space and time. There are some early attempts to parameterize this into process models like Eq. (3.2), though the quantity of enzymes is assumed to be proportional to the microbial biomass, and it is not based on specific genetic information. To summarize, ample data from the field and lab can parameterize Eq. (3.1), while it is difficult – except through new research – to parameterize Eq. (3.2) across the landscape: the focus of this book. Thus, the biology of soil biogeochemistry is implicitly embedded into general constants while recognizing (1) that there is a stunning complexity of life behind it and (2) that during the coming years, there may be an ability to explicitly include the geographical variation of metabolic capacities in these models.

As an example, Bru et al.[23] extracted DNA and used polymerase chain reaction (PCR) amplification to quantify the spatial variation in genes that code for key enzymes in N cycling in the soils of Burgundy, France. It should be noted that this method may be incomplete and miss enzymes, but the emphasis here is on the novelty of the approach. Illustrated in Figure 3.11 are the spatial distributions of the genes *nirK* and *nirS*, genes that code for nitrite reductase and the production of N_2O, and the gene *nosZ*, which codes for nitrous oxide reductase. There is growing interest in how the ratios of these genes (*nosZ/ nirK,nirS*) may ultimately determine whether soils emit N_2O (a strong greenhouse gas) or the harmless form N_2. Ultimately, by understanding and predicting patterns of metabolic capacity, we may be able to target management steps to better reduce the environmental impacts of soil processes.

3.6 Plants

If the Bacteria and Archaea on Earth provide the biochemistry that facilitates biogeochemical cycling in soils, the "loop" in the biogeochemical cycle is really complete once plants are added to the system. Plants are Eukaryotes and only appeared on Earth late in its history. The global proliferation of the Eukaryotes has led eventually to the extensive biotic exploitation of dryland surfaces.

There was soil, and there were soil cycles, prior to the evolution of plants, but the boundary conditions for these processes were certainly different: (1) the atmosphere was low in O_2 for billions of years, leading to enhanced mobility of elements like Fe that today are commonly very insoluble in soil, (2) as a result of low O_2, there was no ozone layer, and life at the immediate land surface was likely extremely difficult, (3) low organic C and N (due to lack of plant inputs) meant that soils had fewer organic compounds, processes, and certainly soil CO_2, and (4) soils were exposed to rainfall and advective erosion processes. On relatively level surfaces, with no or gentle slopes, physical erosion rates may have been slow, so that the weathering and creation of soil may still have occurred. However, on sloping topography, erosion rates would have certainly been higher relative to our present plant-covered planet, and there is speculation that there may have been only a thin soil cover on upland terrains prior to the evolution of plants.[24] Whatever the erosional conditions, there are now more than 30 studies of Precambrian paleosols,[25] and it is clear that many of these Precambrian soils have undergone significant chemical alteration, to great depths, in what was likely a high–atmospheric CO_2 environment. One important consideration with paleosols is possible truncation of the soil surface by erosion prior to burial (possibly removing any "biotic" zone) and the loss of any C and N during burial. Yet, the bulk geochemistry can remain as a signal of the similarities to, and differences from, modern weathering. In modern, oxic soils that undergo significant chemical weathering in warm tropical environments such as Puerto Rico, both Fe and Al are partially lost, but due to the relatively insoluble secondary minerals they form, they tend to be retained in similar proportions. In contrast, in the anoxic or low-O_2 environment of ~2.4 or so billion years ago, at the time when atmospheric O_2 levels are suspected of having begun to rise, Fe once released from silicate minerals was largely in the soluble Fe^{+2} form and was more heavily depleted from soils relative to Al than might be expected today[26] (Figure 3.12).

The eventual creation of an ozone layer reduced surface radiation and certainly created a better habitat for Eukaryotes to colonize the immediate land surface. Colonization is an important way to physically stabilize the soil surface and reduce erosion rates. This in turn can increase soil residence times and the duration of soil exposure to chemical weathering processes prior to its erosive removal. One research report suggests that this instituted a feedback loop in the late Precambrian that greatly impacted the C and O cycles. The authors proposed, based on the increasing apparent abundance of kaolinite and expandable clays in the late Precambrian, that early enhancement of soil processes occurred through the colonization of land by fungi or other early Eukaryotes.[27] Clay is a good template to adsorb and retain organic C, and the eventual erosion of soils (and clay+carbon), and their burial in oceans, created a sink for carbon (a global redox half-cell) and the subsequent increase in

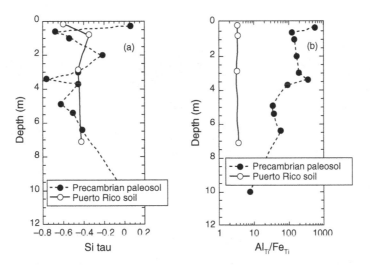

Figure 3.12 A comparison of the geochemistry of a 2.2 to 2.4 By paleosol (present-day Finland) and a modern wet tropical forest soil from Puerto Rico. Both soils formed on granitic lithologies, though the Puerto Rico rock appears to contain more Fe. In (a) the Si tau values (using Ti for paleosol, Zr for modern soil) are similar for both soils, showing 20 to 80 percent loss via weathering. In contrast, the ratio of Al to Fe (both normalized to Ti) shows significant apparent surface losses of Fe relative to Al in the paleosol (note log scale), while in the modern soil both elements exhibit similar patterns of gains and loss with depth. The paleosol formed approximately at the time that atmospheric O_2 levels began to increase (Figure 3.5). Data from A. F. White et al., Chemical weathering in a tropical watershed, Luquillo Mountains, Puerto Rico: I. Long-term versus short-term weathering fluxes, *Geochimica et Cosmochimica Acta*, 62: 209–226 (1998), and S. L. Stafford, Precambrian paleosols as indicators of paleoenvironments on the early Earth, PhD Dissertation, University of Pittsburgh (2007).

atmospheric O_2. Eventually, O_2 increased to levels suspected of supporting larger animals, contributing to the biological evolution of that time.

While it is interesting to investigate, and speculate, about soil formation in a world without plants, the present world provides us with a framework to quantify their impacts on soils and soil biogeochemical cycles. There are three ways plants help or control biogeochemical cycling in soils (Figure 3.13).[28] (a) Plants drive the cycles of C, N, and S – elements that are largely not found in the silicate minerals of soils but are derived from the atmosphere; (b) plant uptake and demand for elements can lower soil solution concentrations below equilibrium values, enhancing rates of mineral weathering; and (c) plant inputs of extracted elements can cause soil solutions, in certain parts of the profile, to exceed equilibrium concentrations and facilitate the formation of secondary minerals. It is likely that (b) and (c) can occur, at different depths, in the same soil.

Plants have multiple effects on soils that enhance or control nutrient cycling:

- The release of organic acids, organic chelating compounds, and CO_2 combines to increase acidity and element solubility and enhance chemical weathering rates.
- Plants are soil water pumps and have vertical variations in root density that control soil water content. Extraction of water, combined in some cases with selective nutrient exclusion by roots, can facilitate secondary mineral formation.

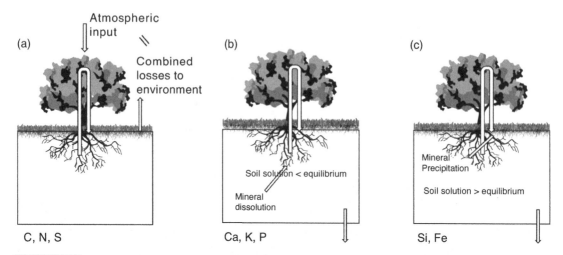

Figure 3.13 Contrasting ways in which plants help control soil chemistry. (a) For elements derived largely from the atmosphere, plants and microbes conspire to create a quasi-steady state, where inputs eventually match outputs. (b) For biologically essential elements found in low concentrations, plants reduce soil solution concentrations due to uptake, enhancing rates of mineral dissolution and nutrient recycling. (c) In some environments – such as some wet tropical soils – plants acquire elements from depth and recycle them near the soil surface, causing high soil solution concentrations and the precipitation of secondary minerals. Modeled from Y. Lucas, The role of plants in controlling rates and products of weathering: Importance of biological pumping, *Annual Review of Earth and Planetary Science*, 29: 135–163 (2001).

- Plants can facilitate mineral formation directly and indirectly. Indirect ways include the concentration of Ca ions and CO_2 around roots, which facilitates the precipitation of $CaCO_3$. In wet, tropical environments, the deposition of biologically cycled Si by leaf litter allows the upper part of soils to maintain the presence of significant quantities of kaolinite when rates of leaching should conceivably favor Si removal and the formation of Al oxides.[29] Plants can directly become mineralized. $CaCO_3$, Ca oxalate, and opal are all minerals that are found in a variety of plants, and biological cycling of Si is now known to be an important part of the global Si cycle.[30]

- Redistribution and retention of elements in soil profiles. The extent, and element-specific nature, of elemental redistribution has recently been compiled by Jobbágy and Jackson.[31] For a soil that is chemically homogeneous with depth at t = 0, one would expect (in an environment without plant cycling) that the elements would become depleted near the surface over time, and the impact of depletion should decline with depth (Figure 3.14b). In contrast, if an element is biologically recycled because it is needed by vegetation, over time the surface of the soil will become relatively enriched relative to the rest of the soil (Figure 3.14a). The authors used the NRCS database (discussed in Chapter 5) to compile global patterns of elemental distribution with soil depth. They found that (a) for C, and the essential nutrients P, N, and K, surface enrichment was nearly ubiquitous (Figure 3.14a); (b) for Ca and Mg, which are less important, the trends with depth were roughly constant; while (c) for elements of little biological importance (e.g. Na and Cl), the surface was relatively depleted (Figure 3.14b). Another important finding was that for soils low in a

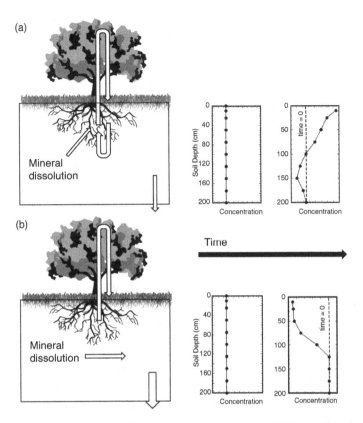

Figure 3.14 Changes in soil element depth patterns over time for two scenarios. (a) Plants efficiently scavenge and recycle nutrients, leading to an enrichment near the soil surface above parent material values and a possible depletion at depth due to root uptake. (b) For elements that are not biologically essential, chemical weathering depletes the concentrations near the surface below parent material values with no subsurface enrichment. Modeled from E. G. Jobbágy and R. B. Jackson, The distribution of soil nutrients with depth: Global patterns and the imprints of plants, *Biogeochemistry*, 53: 51–77 (2001).

nutrient, such as K, the enrichment near the surface increased, indicating strongly biological scavenging and recycling. The study largely focused on exchangeable or water-extractable ions, but the trends should be expected to some degree in the total chemical concentrations of the soils.

As might be expected from the patterns discussed, plants have acquired genetically controlled nutrient regulatory mechanisms that not only promote or inhibit uptake but also can cause the formation of secondary minerals within the plant itself. Some of the most common minerals that form, sometimes abundantly, in plants include opal (SiO_2-H_2O), calcite and aragonite ($CaCO_3$), and calcium oxalate ($CaC_2O_4 \cdot (H_2O)_x$). A recent review[32] identifies some of the important ways in which biomineralization aids plants:

- Calcium regulation
- Heavy metal and toxicant regulation

- Light regulation
- Facilitating the release of pollen and germination
- Herbivore deterrence
- Regulation of soil biogeochemistry

In addition to the chemical role of plants in soil processes, plants exert a critical physical function through soil stabilization and the shielding of the soil surface from rainfall impact. As discussed in later chapters, this helps change the major mechanism of soil movement on hillslopes from advective processes to slower, biologically mediated, diffusive processes. This allows soil thickening, the increase in soil residence times, and the encroachment of biotically rich ecosystems onto sloping land surfaces.

3.7 Biosequences

As discussed earlier, soil science is just beginning to learn how climate and time impact the community and functioning of soil microorganisms, and it can be hypothesized that soils and biota coevolve over time in a complex set of feedback mechanisms. An intriguing theoretical question is how soil and ecosystem processes would change if all factors, except *potential biota,* were constant. In other words, how would soil processes differ if the ecosystem had an entirely different influx of genetic material? This can, of course, be done experimentally over relatively short timescales. An example is the large-scale lysimeter studies that have been established in southern California.[33]

However, nature itself likely provides natural biotic experiments. To find them, one must look for sites that have the same climate, topography, geology, and soil ages – but that are located in geographically separated regions, allowing the influx of a fundamentally different potential biota.

One possible comparison may be that of Mediterranean, highly seasonal ecosystems on the west coast of North and South America. While they are not perfectly identical physically, there are many common factors that can be matched in both locations. However, to a large degree, the potential biotas of the two regions are fundamentally different. Some early, but important, books have been devoted to these questions[34] from the perspective of the ecological and evolutionary questions. To date, the effect of the biotic differences on soil biogeochemistry has not been well investigated.

Yet, the most important question about the relationship between biology and soil biogeochemistry (and Earth surface processes in general) is what soils and the Earth's surface would be like without life. It's a question seldom asked. Addressing the question is fundamentally challenging, in that life has adapted to and proliferated over nearly the entire planet Earth, leaving no "control experiment" with which we can compare in its wake. Some of the approaches that have been made to explore around the edges of this question include:

- Experimental or observational ecosystems with biology and no biology[35]
- The study of soils and ecosystems, of the distant and pre-land plant past, that are preserved in the geological record[36]

- Modeling
- Climosequence studies, at the driest edges of the Earth's environment, to examine how biogeophysical processes change as both water and life disappear[37]

All these approaches are illuminating but also have weaknesses. A seemingly related simple question (does life have a detectable impact on the shape of the Earth's surface?)[38] is also a vibrant research field with no apparent definitive, overarching answer.

3.8 Summary

The emergence of life early in Earth history set in motion an evolutionary trial-and-error process that resulted in a diversity of life we are only beginning to fathom and brought about fundamental changes to the planet on which we live. The evolution of photosynthesis eventually expanded to the point that the chemistry of the atmosphere was altered, new sets of coupled chemical processes involving oxygen and electrons emerged, and novel sets of minerals formed that were not possible in an abiotic environment.

This system, which is collectively called the Earth's biosphere, is considered a complex system, one in which the whole is greater than the sum of its parts. The biological entities, acting in accordance with their individual strategies to survive and reproduce, collectively changed their planetary boundary conditions and in turn, set the stage for the subsequent evolution of differing forms of life.

Soil is a part of this complexity. In Chapter 9, we consider how soil thickness on hill slopes is maintained by, in some locations, two independent negative feedback systems, which may respond to perturbations by driving soil thickness back to its local steady-state conditions. In Chapter 10, we consider how global soil C is (along with microbes) part of a positive feedback loop with atmospheric CO_2 and temperature, a feedback with potentially serious consequences for humanity as we continue to elevate atmospheric greenhouse gas concentrations. It seems likely that many aspects of the biogeocomplexity of the soil system remain to be discovered or recognized, making this one of the most exciting and dynamic fields of soil biogeochemistry.

3.9 Activities

3.9.1 Soil Biology and Role in Modern Medicine

Only one soil scientist has been awarded a Nobel Prize: Selman Waksman of Rutger's University for medicine. Wikipedia, the *New York Times*, and the Nobel website have interesting historical and biographical information on Waksman.

- What was the organism (and its domain of life) that won Waksman fame?

- Where was the organism found that was the center of the Nobel award?
- Where have many of the antibiotics we commonly use been derived from?
- What controversy arose over the Nobel Prize award to Waksman?
- Why is there an urgent need to continue to discover natural sources of antibiotics?

3.9.2 Earthworms and the Biotic Factor

Areas that were glaciated lack earthworms, and the soils and ecosystems have thus evolved there in the absence of worms and their processes. Due to human activities, we have begun to introduce earthworms into areas that formerly lacked them, causing sometimes unexpected changes in soil processes and properties.

Using Wikipedia or other online resources,

a. What are the three functional types of earthworms, and how does each type interact with the soil?
b. How can the introduction of earthworms into an ecosystem be best viewed in state factor theory?
c. In an innovative study, Alban and Berry (1994)[39] published data on a 14-year period of observations of the changes to soil properties driven by earthworm invasion in a forest in northern Minnesota. Plot a best fit model through the data and report the rate of change in the forest floor and surface soil C over time with appropriate units for each rate. Use units of $kgC\, m^{-2}$.

Year	Forest Floor C (Mg/ha)	Surface Soil C (Mg/ha)
1	22.8	13.8
2	25.7	16.1
3	22.3	15.2
4	18.7	20.7
5	–	–
6	7	20.7
7	10	22
8	–	–
9	4	27.5
10	–	–
11	–	–
12	1.7	26.7
13	–	–
14	1.9	28.8

3.9.3 Soil Elemental Profiles and Biocycling

As introduced in this chapter and discussed in greater depth later in the book, chemical weathering moves slowing downward in a soil, releasing soluble cations or anions and removing them if they are not biologically important. In the associated spreadsheet are selected data from a climosequence (elevation gradient) on the island of Hawaii (Porder et al., 2007[40]). As elevation increases, rainfall increases while temperature declines. Combined, these trends ensure that there is greater water and chemical weathering as elevation increases.

Plot the depth trends (for all elevations) for Na, Ca, and P individually.

a. From a web search, how or why is each element essential (or not) to plants?
b. How do the depth profiles conform to expectations based on relative biological requirements? Do any elements exceed that of the parent material (basalt)?
c. How do the depth trends change with elevation, and what are the reasons for this?

4 Field-Based Properties of Soils

> The fascinating impressiveness of rigorous mathematical analysis, with its atmosphere of precision and elegance, should not blind us to the defects of the premise that condition the whole process.
>
> T. C. Chamberlin commenting on Lord Kelvin's calculation of the age of the Earth[1]

The ability to perceive and recognize the features of soils in the field is essential for the appropriate collection of samples for laboratory analyses, as well as for developing hypotheses and appropriate mathematical models of soil processes. Informed field observations and measurements are the important first step in understanding soil biogeochemistry.

Soil profiles, which are two-dimensional vertical exposures of the layering of soils, are examined through hand- or machine-excavated pits or trenches, or occasionally through natural erosional exposures on the landscape. Once a soil profile is accessible, the layers, or *soil horizons*, are identified through a combination of visual and tactile procedures. These observations are recorded and are used as the basis for developing hypotheses about the processes that have shaped the soil's features. This chapter briefly provides an overview of the information recorded on field data sheets and how this information can supplement detailed laboratory measurements. The skills required to carry out a soil examination in the field cannot be adequately introduced in a book; they are best illustrated in the field with an accompanying field course. There is a standard reference for the procedures used in field studies of soils, the United States Department of Agriculture (USDA) Soil Survey Manual,[2] which should be consulted in preparation for fieldwork. The objective in this chapter is to introduce a framework of important physical observations that can be a start to soil biogeochemical research.

Soil field characterizations are semiquantitative, but they are skills that are enhanced, and rendered reproducible, with continued practice. Additionally, skilled sampling is fundamental to reliable subsequent laboratory-based research on soil processes.

4.1 The Soil Profile

The Russian pedologists of the late nineteenth century,[3] while certainly not the first to recognize that soil has vertical variations in properties, were the first to place systematic emphasis on these variations as being an essential defining feature of the soil. These scientists recognized that profile features result from a unique combination of the factors of soil formation: climate, biota, topography, parent material, and time. "This was a revolutionary concept, as important to soil science as anatomy to medicine."[4]

Figure 4.1 An exposed soil profile (in a roadcut), illustrating the "anatomy" of the San Joaquin soil series, the state soil of California. The profile description of a soil much like this is given in Table 4.1.

The key anatomical feature of soils is the vertical layering observed in the field. These layers are the result of a multitude of pedogenic processes and have been termed soil horizons. Soil horizons can be distinct or subtle, thick or thin, caused by biological or almost purely physiochemical processes. For example, the distinctive darkening of upper layers of soil (what are called "A" horizons) reflects the near-surface inputs of organic matter from plants, biological mixing by insects and animals, and the slow downward diffusive transport, combined with biological degradation, of the organic matter. Quantifying these processes and modeling them can be complex (Chapters 7 and 8), but astute and detailed field sampling is needed to detect vertical changes that inform modeling efforts. It is also important to recognize that soil features vary continuously, and commonly

Table 4.1 A profile description of an example of the San Joaquin soil series from California. Identification and explanation of terms are provided for the Ap horizon.

Ap – Horizon Location: 0 to 15 centimeters (0.0 to 5.9 inches); **Soil Munsell color (dry and moist)** brown (7.5YR 5/4) interior **(Soil texture):** loam, dark brown (7.5YR 3/4) interior, **Soil structure:** moist; moderate fine and medium subangular blocky structure; **Soil consistence (dry, moist, moist stickiness, moist plasticity):** hard, friable, slightly sticky, slightly plastic; **Root size and abundance:** common very fine roots; **Pore size and abundance:** many very fine interstitial and tubular and few fine tubular pores; **Special observations:** 2 percent manganese or iron-manganese stains; 1 percent iron-manganese concretions; **Soil acidity:** slightly alkaline, pH 7.5, Hellige-Truog; **Nature of boundary with the next layer:** clear wavy boundary. Lab sample # 84P01598. Assumed source of lime applied as soil amendment; very few iron-manganese concretions concentrations; very few manganese or iron-manganese stains surface features; common very fine roots

Bt1 – 15 to 25 centimeters (5.9 to 9.8 inches); brown (7.5YR 5/4) interior loam, reddish brown (5YR 4/4) interior, moist; moderate medium subangular blocky structure; hard, friable, slightly sticky, slightly plastic; common very fine roots; common very fine interstitial and many very fine interstitial and tubular pores; 2 percent manganese or iron-manganese stains and 15 percent clay films on faces of peds; 1 percent iron-manganese concretions; moderately acid, pH 5.7, Hellige-Truog; clear wavy boundary. Lab sample # 84P01599. Very few iron-manganese concretions concentrations; few clay films surface features on faces of peds; very few manganese or iron-manganese stains surface features; common very fine roots

Bt2 – 25 to 41 centimeters (9.8 to 16.1 inches); brown (7.5YR 5/4) interior loam, reddish brown (5YR 4/4) interior, moist; moderate medium angular blocky structure; hard, friable, moderately sticky, moderately plastic; common very fine roots; many very fine and fine interstitial and tubular and few fine tubular pores; manganese or iron-manganese stains and 15 percent clay films on faces of peds and 15 percent 10YR 7/2) skeletans on faces of peds; moderately acid, pH 5.9, Hellige-Truog; abrupt wavy boundary. Lab sample # 84P01600. Few clay films surface features on faces of peds; few skeletans (sand or silt) surface features on faces of peds; common very fine roots

Bt3 – 41 to 53 centimeters (16.1 to 20.9 inches); brown (7.5YR 5/4) interior clay, strong brown (7.5YR 4/6) interior, moist; moderate medium prismatic structure; extremely hard, firm, moderately sticky, very plastic; few very fine and fine roots and few medium roots; few very fine interstitial and common very fine tubular pores; 7.5YR 3/4) clay films and 15 percent 7.5YR 4/6) clay films on faces of peds and 15 percent manganese or iron-manganese stains; 1 percent iron-manganese concretions; neutral, pH 7.0, Hellige-Truog; gradual wavy boundary. Lab sample # 84P01601. Structure expected to be strong when dried out. Also common pressure faces and in matrix nonintersective slickensides; few iron-manganese concretions concentrations; few clay films surface features on faces of peds; few manganese or iron-manganese stains surface features; few very fine and fine roots; few medium roots

Bt4 – 53 to 66 centimeters (20.9 to 26.0 inches); brown (7.5YR 5/4) interior clay loam, brown (7.5YR 4/4) interior, moist; moderate medium prismatic structure; extremely hard, firm, moderately sticky, very plastic; few very fine and fine roots and few medium roots; few very fine and fine interstitial and common very fine tubular pores; 15 percent manganese or iron-manganese stains and 37 percent clay films on faces of peds; 1 percent iron-manganese masses; neutral, pH 7.3, Hellige-Truog; abrupt wavy boundary. Lab sample # 84P01602. Structure expected to be strong when dried out. Also common pressure faces and in matrix nonintersective slickensides; few soft masses of iron-manganese concentrations; common clay films surface features on faces of peds; few manganese or iron-manganese stains surface features; few very fine and fine roots; few medium roots

Bqm1 – 66 to 74 centimeters (26.0 to 29.1 inches); 80 percent brown (7.5YR 5/4) interior and 20 percent light brown (7.5YR 6/4) interior unknown texture, brown (7.5YR 4/4) interior, moist; massive; extremely hard, indurated by carbonates and silica; common very fine tubular pores; 1 percent fine iron-manganese concretions; strong effervescence, by HCl, 1 normal; moderately alkaline, pH 8.0, Hellige-Truog; gradual smooth boundary.

Table 4.1 (Cont.)

Lab sample # 84P01603. Lime on fracture faces and filling narrow open spaces in some fractures. 90 percent silica and sesquioxide cementation in matrix; few fine iron-manganese concretions concentrations

Bqm2 – 74 to 122 centimeters (29.1 to 48.0 inches); 40 percent brown (7.5YR 5/4) interior and 60 percent strong brown (7.5YR 4/6) interior unknown texture, dark brown (7.5YR 3/4) interior, moist; massive; extremely hard, indurated by carbonates and silica; common very fine tubular pores; 1 percent fine iron-manganese concretions; strong effervescence, by HCl, 1 normal; moderately alkaline, pH 8.0, Hellige-Truog; clear wavy boundary. Lab sample # 84P01604. Lime on fracture faces and filling narrow open spaces in some fractures. 90 percent silica and sesquioxide cementation in matrix; few fine iron-manganese concretions concentrations

Bq – 122 to 152 centimeters (48.0 to 59.8 inches); brown (7.5YR 5/4) interior unknown texture, dark brown (7.5YR 3/4) interior, moist; massive; extremely hard, indurated by carbonates and silica; many very fine interstitial pores; 1 percent fine iron-manganese concretions; strong effervescence, by HCl, 1 normal; moderately alkaline, pH 8.0, Hellige-Truog. Lab sample # 84P01605. Lime on fracture faces and filling narrow open spaces in some fractures. 70 to 90 percent silica and sesquioxide cementation within matrix; few fine iron-manganese concretions concentrations

nonlinearly, with depth. The finer the depth increments of sampling, the greater the information that will be available about a process of interest. In the following sections, the means of identifying, characterizing, and interpreting horizon features in the field are introduced.

4.2 Field Data Sheets

Figure 4.1 and Table 4.1 are a photo and profile descriptions of the San Joaquin soil, the state soil of California (the photo is not of the exact location of the description, but it captures the profile described). The information contained in this table is useful data for a variety of scientific and practical uses. Here, discussions of data having a direct bearing on processes of soil formation is emphasized, and it is shown how these can be linked to chemical or physical properties of soils that can be determined even more quantitatively in the lab.

4.2.1 Soil Color

Most soils are dominated by silicate minerals. A glance at an intrusive igneous rock, like a granite, reveals that the minerals are dominantly (in general terms) clear (quartz), milky (plagioclases), pink (K-feldspars), and black (biotite, hornblende). Yet in soils, these minerals generally play a small role in determining soil color because they are generally coated with varying amounts of organic matter and fine-grained oxide coatings.

Soil color is determined by comparing a soil sample (either in a crushed state or in an intact soil fragment or ped) with a color chip in a Munsell color chart in both a wet and a dry state.[5] The Munsell color system is organized in such a way that an individual page in the book represents a dominant spectral color (hue) or mixture of colors (10YR is a mixture of

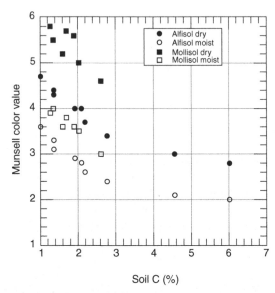

Figure 4.2 Relationship between Munsell color value and soil carbon for both moist and dry soil samples.

yellow and red, for example), the vertical axis on a page represents changes in value (relative lightness or darkness: 0 for absolute black to 10 for absolute white), and changes in the horizontal axis represent variations in chroma (the relative strength of the color, or its brightness: 0 for neutral grays to a maximum of about 20). The correct notation is "hue value/chroma," e.g. 10YR 4/3.

While the color of an individual soil horizon can result from a combination of coloring agents, some general inferences about soil properties can be made from reported Munsell colors.

- *Soil organic matter content*: Soil organic matter (SOM) darkens soil, and an increase in SOM decreases the Munsell color value (Figure 4.2). As an example of the use of this relationship, in the US Soil Taxonomy system of soil classification,[6] values <5 (dry) and <3 (moist) are required for a soil to have a Mollic epipedon, a soil surface horizon characterized by a minimum of 0.6 percent organic C.

- *Iron oxide content*: Iron oxide minerals have their own specific ranges in colors. Goethite has hues ranging from 7.5 YR to 2.5 Y, ferrihydrite from 5 YR to 7.5 YR, and hematite from 7.5 R to 5 YR.[7] Thus, yellowish-brown colors may be indicative of significant goethite contents, while brilliant reds indicate substantial amounts of hematite. The presence of any one of these minerals does not immediately impart these colors to a soil – the amount of the mineral present and the colors of other minerals and organic components are also important. In many locations, the most frequent hues for soils are between 2.5Y and 10YR, the ochres and earth tones associated with soil and weathered rocks, which possess a mix of various oxides and organic compounds.

More generally, soils redden as they age due to increased amounts of chemical weathering and increased concentrations of any of the iron oxides. As weathering proceeds, the hue of the soil becomes redder and, commonly, the chroma of the sample increases. Some researchers[8] have combined the changes in both hue and chroma that soils experience with time into a rubification index, and as expected, there is a pronounced increase in rubification with time.

- *Calcium carbonate content*: Increases in the calcite (white) content of soil causes corresponding increases in the value of the Munsell color.

4.2.2 Soil Texture

Soil texture, as determined in the field, is an estimate of the relative proportions of sand, silt, and clay in a sample (i.e. the sum of particles <2 mm in diameter), usually determined by "the texture feel test."[9] Briefly, in a moistened sample, the stiffness or plasticity of soil increases with clay, and the relative grittiness is indicative of sand content. Modifiers are added to texture names if significant volumes of particles greater than 2 mm in diameter are present. A terminology has been established whereby samples with a limited range in the proportion of the three particle sizes are assigned a textural class name, a strategy illustrated on the standard textural triangle (Figure 4.3). A cursory review of the arrangement of classes on the triangle illustrates that they are not evenly spaced; more classes are

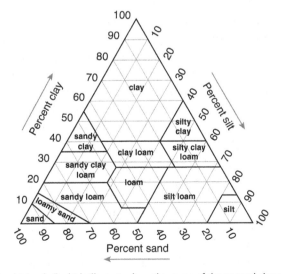

Figure 4.3 The "soil textural triangle," which illustrates how the name of the textural classes of soil correspond to the fractions of sand, silt, and clay-sized particles.

segregated in the lower half of the triangle. The reasons for this are severalfold. First, most common soil textures are found in the lower half of the triangle (i.e. soils dominated by more than 50 percent clay are relatively uncommon in many, but not all, regions). Second, small increases in clay in a soil sample have marked effects on its physical behavior, thus necessitating relatively closely spaced classes in the lower left-hand corner of the triangle. In fact, because of the importance of clay percentage as a guide to processes of soil formation, many field scientists specifically estimate and note the clay percentage of a sample in a separate portion of a field sheet.

Soil texture is a feature of soil that is related to soil age (and, of course, the parent material). As many soils age, clay is produced by chemical weathering and portions of this are subsequently transported downward and concentrated in subsurface horizons. Textural characterization both identifies these layers and provides a semiquantitative guide to the history of that soil.

4.2.3 Soil Structure

The individual mineral particles, which are estimated by the textural characterization, combine into aggregates through the binding action of organic matter, oxides, or clay. The aggregates visible with the unaided eye are known as peds, and the shape, size, and strength of these peds are important soil features for both management and deciphering a soil's history.

Near the soil surface, where plant inputs of C are greatest, organic matter has an important bearing on the nature of soil aggregates. Physical entrapment of sand, silt, or clay particles by fine root systems, or fungal hyphae, may help form the granular units, approximately spherical aggregates generally ranging from <1 to 10 mm in diameter (Figure 4.4), that are commonly found in perennial grasslands. Polysaccharides derived from root exudation or microbial metabolism are effective glues that can bind mineral particles.[10] This may also help explain the occurrence of granules near grass roots, which are known to exude relatively large annual fluxes of organic compounds. Finally, many clay minerals and organic materials are polyanions that are capable of being "bridged" by polyvalent cations such as Ca, Mg, Fe, Al, etc., thus facilitating aggregation.

Deeper in the soil, where clay content increases and organic matter generally declines, platy, blocky, prismatic, or columnar structure may be found. It has been suggested that approximately 15 percent clay is needed to produce any of these structures, and increasing clay content increases both the distinctiveness of the aggregates and their frequency. Shrinking and swelling of the soil mass during wetting and drying cycles is considered responsible for the development of the aggregates.[11] During drying, compressional forces are centered around numerous loci in the soil. It is proposed that blocky structure occurs when lateral and vertical shrinkage forces are similar, while prismatic and columnar structures form when shrinkage forces occur predominantly in a lateral direction. The amount of clay and the depth in the soil seem to also have some bearing on whether prismatic or blocky structure develops in a particular soil horizon. Columnar structure is a special type of prismatic structure that is produced by a well-understood mechanism. In soils with high exchangeable Na percentages, rainwater is

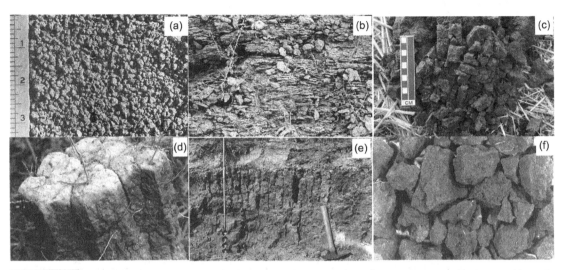

Figure 4.4 An illustration of the shape and relative sizes of different types of soil structure: (a) granular, (b) platy, (c) angular blocky, (d) columnar, (e) prismatic, and (f) sub angular blocky. Photos from the USDA Soil Survey Manual.

capable of dispersing phyllosilicate clay minerals near the soil surface. This causes a downward movement of clays. These clays are then redeposited either through the drying of the soil by evapotranspiration or when the soil solution electrolyte concentration exceeds the critical coagulation concentration (CCC) for the mineral being transported (see Box 4.1).

Soils comprised of a large portion of gravel (>2 mm sized particles) are commonly structureless due to the small, scattered pockets of grains <2 mm. These pockets may be viewed as being too small and discontinuous to contribute to the formation of macroscopic structural units.

It should be emphasized that the presence or absence of macroscopic soil structure does not negate the likelihood that microscopic binding of particles is an important feature of a soil being examined. A given macroscopic soil aggregate may be made of smaller units, which in turn are composed of increasingly smaller aggregates. Thus, it has been suggested that the size distribution of soil aggregates, and their properties, may be described by fractal models.[14]

4.2.4 Clay Films

Clay movement near the soil surface, and its redeposition at a lower depth, during water transport in soil is an important process whose effects become magnified with time. The suspension of clay in downward-moving water can occur through physical forces of the moving water, which dislodges clay from the soil aggregates, or through chemical dispersion caused by low electrolyte concentrations (Box 4.1). Redeposition occurs as the soil

Box 4.1	Phyllosilicate Dispersion and Flocculation

Dispersion and flocculation are important processes affecting secondary silicates that have a bearing on natural processes of soil formation as well as the management of soils during irrigation. Much research has been carried out to determine the ways that irrigation water composition affects clay mineral dispersion in an effort to improve rates of water infiltration into soil. The result of this (and other) research has shown that each phyllosilicate mineral has its own *critical coagulation concentration* (CCC) for a given cation (or combination of cations) in the surrounding soil solution. The CCC is the concentration at which a suspension of minerals (platelets dispersed in solution due to the repulsive negative charges of each platelet) will become attracted as a result of the presence of a sufficient quantity of interlayer cations. Of special importance are the relative effects of Ca^{+2} and Na^{+1} on the flocculation or dispersion of clay-sized minerals, because these are two cations that dominate the positive charge of many soil waters. Calcium is a divalent ion, with a relatively small hydrated radius (in solution, cations surround themselves with a shell of water molecules). A relatively low concentration of Ca in solution is sufficient to counterbalance the net negative charge of phyllosilicates, allowing adjoining platelets to become associated with one another by sharing the interlayer divalent cations. In contrast, it requires a much larger concentration of Na (with its monovalent charge and large hydrated radius) to achieve the same result. The addition of a small amount of Na to a Ca-rich solution has a pronounced effect on the dispersibility of the solution, and a much higher concentration of cations is needed for each additional increment of added Na to prevent dispersion or cause dispersed clay to flocculate.

The concentration of cations needed to flocculate smectite and illite is a function of the relative proportion of Ca vs. Na in the solution. The ratio of these two ions is reported as the *sodium adsorption ratio* (SAR) = $[Na^+]/([Ca^{+2}+Mg^{+2}]/2)^{0.5}$, a parameter of the solution chemistry that has been found to be roughly proportional to the ratio of Na to Ca adsorbed onto the clay. In practice, it has been found that when soil solutions with SARs of 13 or more are passed through soils, numerous detrimental effects occur (swelling, dispersion, and reduced irrigation water infiltration rates).

Under natural soil-forming conditions, the SAR of soil water plays a prominent role in soil characteristics. For example, in soils situated slightly above a high water table in arid environments, long-term evaporation of the ground water during hot dry periods of the year will concentrate Na salts near the soil surface.[12] During the brief rainfall periods, relatively pure and dilute rainwater falls onto the soil surface, reacting with the highly soluble Na salts and producing high-SAR soil solutions, which (if dilute enough) are effective in dispersing clay minerals. As this water passes downward through the soil and increases its concentration of Na salts, the flocculation value of the clays may be reached and redeposition of the clays may occur. This process is responsible for extensive acreages of Na-rich soils, with prominent clay-enriched subsurface horizons, that ring the former marshlands of the Great Valley of California (Whittig and Janitzky, 1963) and which are found around former pluvial lake beds in the Mojave and Great Basin deserts of western North America.[13]

water is lost through evapotranspiration or as the electrolyte concentration increases. The redeposited clay is located along soil features where the water moves: the walls of pores and on ped faces. Clay films are recognized using a standard 10× hand lens. Films display textural and/or color contrasts to the underlying soil matrix. Identification is not always easy, and recognition of films varies with lighting and moisture content. The development of clay films and a subsurface horizon enrichment in clay are features that require periods of time on the order of 10^3 to 10^4 years in many settings. Thus, clay films are important features indicating a substantial time for soil formation.

4.2.5 Carbonate

The presence or absence of carbonate in the soil is determined by the soil's reaction with 1 N HCl: visible effervescence indicates the presence of carbonate, and the degree of foaming is proportional to the amount of carbonate present. Carbonate is a relatively soluble mineral in soils, and the presence, absence, and vertical distribution of this mineral are important guides to the depth of water movement and the soil's age.

4.3 Soil Horizon Nomenclature

An important component of field data sheets that has not yet been discussed is the names of soil horizons – e.g. "A," "C," etc. The reason why we consider horizon nomenclature following the discussion of other field features is that soil horizons are named after the collection of all pertinent data, because the terminology chosen represents the field scientist's estimate of how each soil horizon differs from its starting material. Viewed somewhat differently, *horizon nomenclature represents a hypothesis about processes responsible for the formation of that horizon.* The term chosen to represent an individual horizon commonly requires considerable time and thought and may involve several iterations as more information, especially laboratory data, becomes available and ideas and concepts become clearer.

   Table 4.2 is an abbreviated guide to the nomenclature and definitions of horizons. As the table illustrates, there are six different upper-case terms called master horizons, and a series of lower-case modifiers that may, or may not, be used to add additional information to the master horizon designations. The detailed rules for the use of these terms is found elsewhere.[15] Here we will briefly review the general concepts of the master horizons and discuss a few select uses of modifiers that are commonly added to indicate results of important processes.

   In all cases of naming horizons, the central question that should be asked is "How do the properties of this horizon differ from those of the parent material or the nature of the soil at time = 0?" In soils formed from homogeneous parent material, the nature of the horizon in question can be compared with a sample of the relatively unaltered parent material obtained from the base of the soil exposure. In the case where the parent material varies with depth (e.g. loess over fluvial gravels), more thought must be devoted to assessing the changes to the most surficial layers. The brief definitions of master horizons and lower-case modifiers given in Table 4.2 are expanded in more detail in the standard references for horizon nomenclature prepared by the National Resource Conservation Service (NRCS) of the USDA.[16]

Table 4.2 An abridged listing of the Master Horizons of soil and their definitions

Master Horizons	Definition and Examples of Lower-Case Modifiers
O	Layers dominated by organic matter. State of decomposition determines type: highly (Oa), moderately (Oe), or slightly (Oi)[a] decomposed.
A	Mineral horizons that have formed at the surface of the mineral portion of the soil or below an O horizon and show one of the following: (1) an accumulation of humified organic matter closely mixed with minerals or (2) properties resulting from cultivation, pasturing, or other human-caused disturbance (Ap).
V	Mineral horizon, generally under a desert pavement, that consists of fine-grained eolian material dominated by vesicular pores.
E	Mineral horizons in which the main feature is loss of silicate clay, iron, aluminum, or some combination of these, leaving a concentration of sand and silt particles.
B	Horizons formed below A, E, or O horizons. Show one or more of the following: (1) illuvial[b] concentration of silicate clay (Bt), iron (Bs), humus (Bh), carbonates (Bk), gypsum (By), or silica (Bq) alone or in combination; (2) removal of carbonates (Bw); (3) residual concentration of oxides (Bo); (4) coatings of sesquioxides[c] that make horizon higher in chroma or redder in hue (Bw); (5) brittleness (Bx); or (6) gleying[d] (Bg).
C	Horizons little affected by pedogenic processes. May include soft sedimentary material (C) or partially weathered bedrock (Cr).
R	Strongly indurated[e] bedrock.
W	Water layers within or underlying soil.

[a] The symbols in parentheses illustrate the appropriate lower-case modifiers used to describe specific features of master horizons.

[b] The term illuvial refers to material transported into a horizon from layers above it.

[c] The term sesquioxide refers to accumulations of secondary iron and/or aluminum oxides.

[d] Gleying is a process of reduction (caused by prolonged high water content and low oxygen concentrations) that results in soil colors characterized by low chromas and gray or bluish chromas.

[e] The term indurated means strongly consolidated and impenetrable to plant roots.

4.4 Synthesis of Data

To understand how these properties can be used to create an integrated perspective of a soil's properties and identify some major processes that have affected it, the data in Table 4.1 are used to create a mental and visual image of the soil profile it reflects (Figure 4.5). To begin, the San Joaquin soil forms on granitic alluvial fans in the San Joaquin Valley of California. The fans are largely glacial outwash from the Sierra Nevada to the east and are largely in the 200 to 300 Ky age range. The San Joaquin Valley has hot summers (mean annual temperature (MAT) ~15 °C) and modest mean annual precipitation (MAP) (300 to 400 mm) that falls entirely in the winter. Virtually no rain falls between May and November.

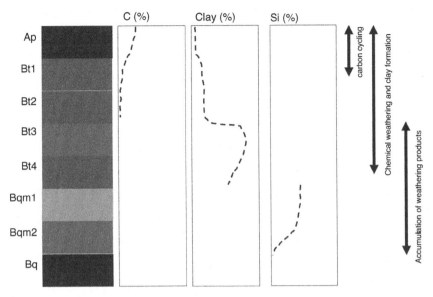

Figure 4.5 A schematic representation of soil properties and processes that shape the San Joaquin soil, based on data from Table 4.1.

The granitic suite of minerals (biotite, hornblende, Ca and Na feldspars, and quartz) undergo chemical weathering in proportion to their relative susceptibility based on their chemistry and structural makeup. Due to the semiarid conditions (and extreme seasonal rain), chemical weathering products are not completely removed, and in particular, Si released from primary minerals accumulates as opal (SiO_2 + H_2O), creating an indurated set of horizons (Bqm1, 2) below the zone of clay accumulation (Bt1 through 4). The flora is an annual grassland, where European grasses have largely replaced preexisting grasses and forbs. The relatively shallow rooting depth and modest annual production of biomass are reflected in the dark colors found only in the Ap horizon (which has likely lost C due to cultivation).

The long duration of weathering, even in a semiarid environment (which likely experiences more rainfall during glacial epochs), has made the upper three horizons somewhat acidic (pH <6, though the Ap has had $CaCO_3$ added). The buildup of clay has greatly enhanced the soil water-holding capacity. Thus, oxidation-reduction processes form notable stains and concretions of Mn/Fe oxides throughout much of the soil. A distinguishing feature of this soil is the abrupt increase in clay from the Bt2 to the Bt3 horizon. The high clay Bt3 has prominent prismatic soil structure. A hypothesis for this abrupt clay increase is that as the soil has gradually become indurated with SiO_2 in the lower horizons, clay has accumulated over this impervious layer. In addition, minor amounts of $CaCO_3$ have accumulated in the Bqm horizons, a mineral that does not form in well-drained soils of this region of California due to the winter leaching that occurs.

Figure 4.6   The spatial distribution of the San Joaquin soil in California. From UC Davis California Soil Resources Laboratory (https://casoilresource.lawr.ucdavis.edu/see/).

The San Joaquin soil, due to the wide distribution of glacial deposits in the Great Valley of California, has one of the largest areal extents of any soil in the state (about 200,000 ha) (Figure 4.6). From the field-based data we have examined, it would seem that these soils would pose significant challenges to farmers in this area, who produce about 20 percent of the nation's agricultural products. Indeed, soils on younger alluvial deposits, lacking restrictive layers, have long been preferentially used for irrigated agriculture and are still the highest-productivity soils today. However, as new lands for farming became scarcer, innovative methods have evolved over the century to essentially homogenize the horizons of the San Joaquin. First, explosives were used to create a rooting zone for orchards. But with the development of high-powered field implements, today, deep plowing with shanks that are up to 2 m long serves as a way to fracture and mix the Bqm and Bt horizons. In Chapter 10, some of the effects, and the rates of processes, of this management on soil biogeochemistry are examined.

4.5 Summary

This chapter comprises a brief introduction to the collection and interpretation of data on soil properties in the field. The goal has been to introduce the methods and terminology so that readers of field data can begin to develop concepts of soil properties and processes that have affected soils. Developing the skills to produce reliable field data of soils requires field learning opportunities, but the ability to interpret collected field data is a worthy goal in itself and one that can begin through reviewing this chapter and the standard sources for this information.

One of the great attractions of soil biogeochemistry, for many students and scientists, is the allure of field observations and collections of soils. This activity, when done within a theoretical framework that drives the selection of sites to address well-framed hypotheses, is a powerful skill and research tool. Being able to "read the landscape," and being able to mentally partition it via the state factor framework and other important models, is rapidly advancing the field. Knowledge of field activities and terminology is also important even if one is a "data consumer" of the vast and growing amount of soil information that is available. From understanding how, and why, these data were collected, better and more reliable meta-analyses will result.

4.6 Activities

4.6.1 See a Local Soil

Regardless of the setting and where you live, you are living on a soilscape, which should be accessible for observation in some manner. First, identify the parent material of the location and use this as a frame of reference for the soil and its features. An ideal and simple classroom approach is to use a soil auger to excavate a core and sequentially place the auger bites on a tarp or cloth. Alternatively, using a shovel, a shallow excavation, with a clean vertical face, can be prepared. If you have access to a Munsell color book, identify the colors with depth and consider how the hue, value, and chroma change and the underlying soil properties that cause this (at the time of this writing, a free iPhone Munsell viewer is available: "Munsell Viewer"). Even without a Munsell chart, one can qualitatively describe the color changes and consider the changes with depth. Using water from a squirt bottle, take a small portion of each soil horizon, wet it up, and work it around in your hand. Rub it between your fingers. If it is gritty, look closely (preferably with a hand lens), examine the minerals and their color, and try to identify which groups they belong to. Is the sample sticky and plastic, and how do these features change with depth? If there is gravel, are the gravel fragments fresh in appearance, or are they soft and weathered? Either way, consider the nature of the rock minerals, and also the approximate amount of time the soil has been forming, as interpretative aids. If you have access to a portable pH meter, or even more simply, pH strips, put a small amount of soil in a plastic cup, add ~2× the volume of water, stir, let stand, and then measure the pH of the solution. If it is basic, neutral, or acidic, what are the parent material and climatic factors that impact this?

4.6.2 See Soils of the World

The World Soil Museum in the Netherlands is a resource for a virtual tour of the diversity of global soils: https://wsm.isric.org. By following the links to "explore collection," soil profiles from around the world can be accessed and examined (virtually). The collection is accessible by nation or by soil classification. Clicking on the profile uploads a detailed soil profile, while clicking on the name brings up data on the nature of the soil – including, in many cases, lab data. Examine the depth changes in color of the Chernozems (grassland soils of the steppes). Examine the depth profiles of the Podzols. From colors (and other online information), attempt to assign soil horizon nomenclature to these vs. depth (or to profiles assigned to you by your instructor) (note: the nomenclature used by the World Soil Museum is similar, but not always identical, to the US system briefly introduced in this chapter). Plot selected data vs. depth (e.g. %C, $CaCO_3$, clay, etc.) and compare with the examined profile and with the horizon nomenclature. If it is reported, explore the mineralogy of the clay-sized fraction of the soil and consider this with respect to depth and the local environment.

It is important to emphasize that in most cases, these soil profiles represent the combined effect of biological additions and mixing near the surface, and the downward migration of water and associated chemical weathering and transport with depth. How strongly these processes are expressed is related to the climatic conditions of the site and the amount of time that has elapsed.

4.6.3 Access Web-Based Soil Tools

Cell phone technology makes enormous amounts of information accessible to a scientist even in the field. This includes:

1. SoilWeb: This mobile device app from the University of California, Davis, is a soil survey in your phone (for the United States). The UC Davis app links to the NRCS national soil database. At any location one may access the soil, a schematic representation of the soil, and profile descriptions (discussed in this chapter) and laboratory data (discussed in the next chapter). For example, download the app, and for your local soil, how does it compare with the information provided by SoilWeb?
2. SoilInfo App: This app is provided by the ISRIC (International Soil Reference and Information Centre) to provide global soil data to the user.

These apps (and likely more that will be developed over time) are essential tools for any soil biogeochemist, providing immediate information on the nature of the soil profile and its entire set of physical and chemical features. When used frequently and with a deliberate plan, they help one develop a geographical understanding of the soil continuum across a given area or transect.

Soil Biogeochemical Measurements and Data

As pedology has evolved along with advances in chemistry and physics, a considerable array of soil properties are now measured on a routine basis. These properties can be divided into two categories: (1) total chemical analyses (which were considered in Chapter 2) and (2) extractive chemical and associated physical analyses. The later large suite of analyses is performed routinely by the Natural Resource Conservation Service (NRCS) during soil survey activities, and there is now an enormous database for the USA that is accessible to address many questions and problems. These data are called *soil characterization data* by the NRCS and can be easily and freely accessed at their website.[1]

The soil properties presently measured reflect not only advances in instrumentation and methods but also evolving concepts about what soil attributes are important for one to develop a basic understanding about soils and how they can be effectively managed. These properties can be used to accurately classify the soil, determine its suitability for various land uses, and help determine it origin and its functioning in biogeochemical cycles.

The organization within the NRCS that produces and compiles soil data in the USA is the Kellogg Soil Survey Laboratory (KSSL) of the National Soil Survey Center in Lincoln, NE. In order to use soil chemical and physical data in problem solving, which is the theme of this book, it is important to understand how the measurements are made, what they signify, and how they can be further manipulated to address specific questions. This chapter is an abridged introduction to these issues, and references are made throughout the chapter to original sources that will provide more information to the interested user.

5.1 Soil Characterization Data

Thousands of soils have been chemically and physically analyzed in the USA. They are all accessible through the NRCS Soil Characterization Data Portal: https://ncsslabdatamart.sc .egov.usda.gov. Also, as discussed in the Activities section of Chapter 4, the data can be obtained on a mobile device via SoilWeb.

Here, one is examined to serve as a representative example. To continue the initial exploration of soil biogeochemistry along the Mississippi corridor (Chapter 1), the soil chosen is the Memphis series, which also happens to be the predominant soil at Graceland, Elvis Presley's home in Memphis, TN (Figure 5.1).

Memphis soils have formed on loess that has been mantled over a preexisting rolling topography. Due to warm temperatures and a humid climate, the soil has undergone noticeable alteration during its postglacial development period (discussed in greater length

Figure 5.1 Historical photograph of Graceland prior to the subsequent development of the property for tourism. The house, about 15 km east of the Mississippi, resides on a significant accumulation of postglacial loess. A representation of a typical profile of the Memphis soil is illustrated to the left.

at the end of this chapter). This location offers an excellent opportunity to learn what characterization data can reveal about a soil. The information behind the following discussion comes from the Soil Survey Laboratory Information Manual,[2] the official resource for better understanding soil data reports like that in Table 5.1.

5.2 Soil Characterization Data Reports

Methods and instruments to measure soil properties change over time, and multiple methods may exist to determine a given soil property. Therefore, the method used to determine each type of data in an NRCS soil characterization report is indicated by an alphanumerical code. This code can then be used to locate it in the Laboratory Manual and determine more about the method and its meaning and limitations (if any).

NRCS data reports are arranged by tiers, which contain a series of rows for each soil horizon, and columns (a specific analytical result for each horizon). For the Memphis soil examined here, the tiers encompass the following types of information:

- Tier 1: Particle size determination (PSDA) and rock fragments
- Tier 2: Bulk density and moisture
- Tier 3: Water content
- Tier 4: Carbon and cations

- Tier 5: Cation exchange capacity (CEC) and bases
- Tier 6: Salt
- Tier 7: pH and carbonates
- Tier 8: Phosphorus
- Tier 9: Clay mineralogy
- Tier 10: Sand and silt mineralogy

The report arrangement and the analyses used vary for soils from various parts of the world and the questions asked by the users in these areas. The arrangement for the Memphis soil is not the same as for all other soils in the database. For example, some soils have been subjected to total chemical analyses, which will be present in an additional tier. Many soils may not have detailed mineralogical analyses. However, Memphis soil is sufficiently general that it serves as a guide to this important information. Next, the information contained in each tier, and its significance, is examined as an introduction to utilizing these data to address environmental and pedological questions. The reason for doing this is that there is no comprehensive introduction to these soil characterization data sheets in any standard soil science text. This background is therefore useful to begin to utilize these data sets for elementary problem-solving exercises. For brevity, not every item on the example data sheets is explained, but one can learn more by accessing the Laboratory Manual referenced earlier. The data report for the Memphis soil considered here is found in Table 5.1.

5.2.1 General Information

The headings at the top of NRCS characterization data reports contain information about the soil that was sampled. Included in these data are: (1) county and state of origin, (2) date of sampling, (3) laboratory name and location, (4) client-assigned lab number, (5) pedon (a technical term for soil profile) name and United States Department of Agriculture (USDA) classification, (6) project file number assigned by Soil Survey Lab, (7) Lab-assigned numbers, and (8) general methods of sample preparation and data reporting (alphanumerical codes, which are outlined in detail in a supplementary resource: Soil Survey Laboratory Methods Manual[3]). In the following sections, the reader should refer to Table 5.1 to observe the data and methods being discussed.

5.2.2 Tier 1. Particle Size Analyses

Particle Size Data

The quantification of the size distribution of minerals and rock fragments in a soil is termed particle size analysis (PSA). Particle size is determined on the mineral fraction (i.e. the nonorganic fraction) of the soil. Organic matter and finely dispersed cementing agents (such as oxides) are generally chemically removed prior to PSA.

Sieving is usually done to remove mineral and rock fragments larger than 2 mm in diameter. The distribution of particles in this fraction is found under "coarse fractions." The rock and mineral fragments <2 mm in diameter (the "fine earth fraction") are broken into sand, silt, and clay-size fractions through a combination of sieving and sedimentation methods. The data in all these analyses are reported on a percent weight of soil basis. An important point to recognize is that most of the subsequent chemical and physical analyses in Table 5.1 are conducted on the <2 mm fraction (sometimes called the fine earth fraction). The rationale for this is that the <2 mm fraction contains nearly all the organic matter and chemical weathering products in soils, and the gravels are considered to be largely inert. Thus, when calculating the mass of specific soil properties per unit soil volume, the volume of the gravel must be removed to arrive at an accurate calculation. This is discussed later in the chapter.

From the data in Table 5.1, we can quickly determine a few things about the soil and its history. First, note the very high silt content of the soil (>80 percent). Such a high silt content is characteristic of eolian deposits and conforms to the report that the soil forms from loess. Second, there is a systematic change in clay with depth: lowest at the surface, a "bulge" between 38 and 137 cm, and then a decline. Assuming that the parent material was homogeneous during deposition, the development of this Bt horizon is a function of downward clay migration and *in situ* chemical weathering of the primary minerals to produce secondary silicates. Finally, there is a sharp decrease in silt, and an increase in sand, at 200 cm, likely marking the contact between the overlying loess and a different material below.

5.2.3 Tier 2. Physical Properties and Cation Exchange

Bulk Density

One of the most important properties of soils, if raw laboratory data (commonly determined on a mass/mass basis) are to be utilized for many biogeochemical problems (where a mass/volume measure is required), is the soil's bulk density. Bulk density is the mass of total soil per unit volume (g cm^{-3}). Soil, unlike the rock it is derived from, is highly porous, so bulk densities of soil are usually far lower than that of common rock, which lacks significant porosity (~2.6 g cm^{-3}). The bulk density reported in the data sheet used here is for the < 2 mm fraction of the soil. Box 5.1 explains how bulk density is used, along with a chemical property, to calculate mass per volume.

Bulk density is not a static soil property – it can vary appreciably with moisture content, particularly in soils with substantial quantities of expandable 2:1 phyllosilicate clays. Therefore, bulk density can potentially be measured in a range of moisture contents (field moisture at time of sampling, after samples have been equilibrated to 1/3 bar of pressure, or at so-called "field capacity" – the amount of water stored after a saturated soil has had sufficient time to drain and oven-dry). As the data in Table 5.1 indicate, bulk density of this soil increases as the water content decreases, particularly in the clay-rich Bt horizons. The changes in bulk density from moist to dry are used to determine the coefficient of linear extensibility (COLE), a measure of the shrink-swell capacity of the soil.

Water Content and Retention

The ability of soil to retain water depends on its particle size distribution, mineral composition, and organic matter content. In addition, from a plant perspective, it is also important to determine the amount of water in the soil between various potentials: one end member being "field capacity" (33 kPa) and the other being "permanent wilting point" (which is roughly 1500 kPa of pressure). The KSSL determines these values by wetting soils and then dewatering them by pressure equilibrations.

The water retention data is traditionally determined for agronomic purposes: it allows irrigators to determine the approximate amount of irrigation water to apply to bring a given depth of soil to field capacity. The amount of water that a soil horizon can hold is given by the water retention difference between different potentials. Additionally, the lab sheet provides water retention difference (WRD), which is in units of water held per unit of soil (length or volume). For example, the Memphis soil has approximately a WRD of 0.25, which means that the entire profile (to 200 cm) can hold roughly 50 cm of water.

Cation Exchange Capacity and Ratio to Clay

As discussed in Chapter 2, CEC is the quantity of cations adsorbed by soil minerals and organic matter and is an indication of the soil's net negative charge at some defined pH. Commonly, the SSL reports CEC determined by two methods: (1) sum of adsorbed cations and (2) displacement of an adsorbed cation by NH_4OAc buffered to pH 7. The first method can overestimate CEC if the soil sample being analyzed has soluble salts or carbonates, which dissolve and give the appearance of a large quantity of adsorbed cations. The second method is arguably the standard method used by many researchers. The results of both methods are reported as cmol of + charge adsorbed per kg of total soil (including organic matter).

For the Memphis soil, the CEC has a trend with depth that roughly mirrors the clay trend with depth, 6–8 cmol charge kg^{-1} soil at surface, increasing to 16, and then declining with depth.

Shown in the "Bulk Density and Moisture" row is the "ratio/clay" of the CEC. As discussed in Chapter 2, the ratio of CEC (determined for total soil) normalized to the clay fraction (where most CEC originates) is a reasonable guide to the mineralogical composition of the clay fraction. The units in Table 5.1 are cmol (+) charge/kg clay, which can be multiplied by 100 to compare with CEC ranges for minerals and organic matter given in Table 2.2. As a guide, the KSSL suggests the following approximations based on soil analyses from the central USA and Puerto Rico:

CEC/%clay	Mineralogy as determined by XRD or DTA
>0.7	smectitic
0.5–0.7	smectitic or mixed
0.3–0.5	mixed
0.2–0.3	kaolinitic or mixed
<0.2	kaolinitic

When we later examine the X-ray diffraction analyses of the clay fraction, we will see that this simple rule of thumb correlates well with more precise measurements.

5.2.4 Tier 4. Chemical Analyses

Organic C and N

The organic fraction of soils is characterized by C and N analyses. Organic matter is about 58 percent C with the remainder being O, H, N, etc. C:N ratios for soils range from values of 30 or more for surface litter layers composed of fresh plant residue to values lower than 10:1 for well-decomposed organic matter in deeper soil horizons. Methods to measure C and N by the USDA vary between wet chemical techniques for C and combustion for N and C.

For the Memphis soil, the landscape is presently being used for farmland (Ap horizon), and the C content of the Ap horizon is modest (<1 percent) and declines nonlinearly with depth. Below the Ap, there is a sharp decline in the AE horizon, an increase in the Bt1, and then an irregular decrease with depth. The presence of an E horizon (bleached color) is indicative of some process that has removed coloring agents and thus some C. As discussed in Chapter 7, the roughly exponential decline in C with depth is due to plant inputs subjected to degradation and downward vertical transport. The N content is given only for the upper three horizons. The C/N ratio (light orange) reveals a very low C/N ratio (10 or less) indicative of both highly biologically decomposed organic matter and the possible presence of NH_4^+ adsorbed to the clay (which provide nonorganic forms of total N).

Extractable Metals

As discussed in Chapter 2, soils may contain appreciable quantities of secondary oxides of Fe, Al, and even Mn as a result of weathering. Chemical extractants have been developed that are capable of dissolving a suite of these oxides without appreciably affecting non-oxide minerals. The dithionite-citrate extraction dissolves and removes most secondary Fe oxide minerals and the organically bound Fe in many soils.[4] In addition, it removes organically bound Al and Al associated with the Fe oxides as well as easily reducible Mn. It should be noted that these extractants are not entirely capable of removing all desired oxides, nor are they always completely unreactive with silicates and other mineral groups.

For the Memphis soil, there is a marked increase in extractable Fe within the Bt horizons, the point where the soil color becomes markedly redder (see Figure 5.1) and where clay has also accumulated.

5.2.5 Tier 5. Adsorbed Ion Chemistry

Extractable Bases and Acidity, Base Saturation

The net negative charge of soil minerals and organic matter is balanced by positively charged ions. The two alkali metals Na and K, and the two alkaline earth metals Ca and Mg, are collectively referred to as "exchangeable bases." "Exchangeable acidity" refers almost entirely to Al^{+3}, which hydrolyzes to $Al(OH)_3$, producing protons. The SSL determines exchangeable bases and acidity by equilibrating a soil sample with NH_4OAc (ammonium acetate) at pH 7.

Base saturation percentage = (exchangeable bases/CEC) × 100. Base saturations lower than 100 percent are presumed to be due to the presence of exchangeable acids. Apparent base saturations >100 percent can arise in soils that have appreciable levels of soluble salts or carbonates (i.e. arid and semiarid soils), which inflate the apparent value of exchangeable bases. In general, as soils weather, and the primary minerals that are the source of base cations are depleted, the base saturation declines and exchangeable acidity increases.

The exchangeable Na percentage is commonly reported separately due to its important role in determining the dispersibility of phyllosilicate clay minerals in soil solutions. Values greater than 15 percent are considered detrimental in the sense that dispersion of clays, and reduced water infiltration rates, may occur if low-ionic-strength water is applied to the soil. The Memphis soil, due to its humid climate conditions and annual leaching of the soil, has less than 100 percent base saturation but still retains a considerable reservoir of cations (60 to 75 percent). Ca is the dominant cation with lower amounts of Mg. Na (due to high water flow through) is very low, and the exchangeable Na percentage is generally 1 or less.

5.2.6 Tier 6. Water-Soluble Chemistry

Water Extracted from Saturation Paste

Commonly for soils from arid and semiarid regions, the water from a saturation paste is removed by suction and the chemistry of the equilibrated water is determined. The value of this measurement is to provide a view of the chemistry of soil solutions, which in turn can be used in thermodynamic studies, geochemical modeling, and agronomic assessments.

The electrical conductivity of the extracts verifies the presence of significant quantities of soluble salts. Electrical conductivity (reported as $dS\ m^{-1}$) increases as the dissolved solid content of a solution increases. As total salinity (and electrical conductivity) increases, the ability of plants to utilize soil water decreases due to an increase in the osmotic potential. Soils with an electrical conductivity (EC) greater than $4\ dS\ m^{-1}$ are referred to as saline soils, although in reality, different plants have differing abilities to deal with dissolved salts in the soil solution, and some salt-sensitive plants are affected at ECs between 2 and 4.

The sodium adsorption ratio (SAR) of the saturation paste is a guide to the exchangeable Na percentage. Soils with SAR >13 are considered to be significantly affected by Na and are referred to as sodic soils.

5.2.7 Tier 7. Soil Chemical Analyses

Soil pH

The pH of soil is dependent on the conditions and methods of measurement. The value obtained varies with whether the measurement is made on a soil–water mixture or water extracted from soil, and the chemistry and amount of solution in contact with the soil. Some common methods of pH measurement include those on the example data sheet: (1) saturated paste pH, (2) pH of a 1:2 0.01 M $CaCl_2$ solution equilibrated with soil, and (3) a 1:1 water to soil solution. A saturated paste is prepared by slowly adding distilled water to dry soil (and mixing) until free water just begins to appear. The other pastes are prepared on the prescribed weight basis. The measured pH of these methods differs in the following general way: the greater the water to soil ratio, the higher the measured pH, and the addition of $CaCl_2$ results in a lower measured pH than for the pure water mixtures.

Soil pH is loosely correlated with base saturation: as base saturation decreases, pH also decreases. Soil pH lower than 3 or 4 suggests the presence of a strong acid such as H_2SO_4 or HNO_3. Soil pH in the range of 4 to 6.5 indicates the presence of exchangeable Al (and acidity). Soil pH from 6.5 to 8 usually indicates base saturations between 75 and 100 percent. Soil pH from 8 to 8.5 indicates 100 percent base saturation and the presence of $CaCO_3$. Soil pH of 9 or more indicates the presence of Na carbonates and a very high exchangeable Na percentage.

For the Memphis soil, saturation extracts were not analyzed. This absence of saturation extract analyses is common for humid soils, which generally have very low solute concentrations in pore waters. Again, this analysis becomes more common, and important, in the more arid regions of the USA. The pH of the Memphis soil (for the $CaCl_2$ method, 1 part of soil to 2 parts water) is between 4 and 5, and as expected, there is about 2 to 4 percent exchangeable acidity in this soil.

Calcium Carbonate

Carbonate minerals (almost exclusively calcite in many soils) are moderately soluble minerals, and their presence and depth distribution are indicative of the amount and depth of long-term water movement. The SSL measures total carbonate by reaction of the soil with acid and then measuring the CO_2 yield manometrically.

For the Memphis soil, the NRCS did not even measure $CaCO_3$ due to its obvious absence. However, in the more arid regions of the western USA, carbonate analyses will be commonly reported in laboratory analyses.

5.2.8 Tier 8. Phosphorus

Extractable P

A soil measurement of interest for ecology and agronomy is the content of P, which is commonly only second to N as a plant-limiting element in terrestrial ecosystems. Because P

Box 5.1	Calculation of Mass per Volume of Biogeochemical Properties

Most chemical data from soils are reported on a mass/mass basis (%, ppm, etc.). These are also known as *intensive* properties of soils, properties that do not depend on the amount or mass of the soil. These data are themselves important and useful for many questions. However, in a growing number of environmental issues, it is critical to determine *extensive* soil properties, which depend on the mass and volume of the soil. As an illustration, knowing the average organic C percentage of global soils (an intensive property) is not helpful in quantifying soil organic C in the global C cycle. To do that, we need the total mass of C, an *extensive* property. In most cases, by defining the volume of the soil and knowing its bulk density (and gravel content), intensive quantities can be converted to extensive quantities for specific needs and purposes.

To convert percent C (or any intensive property) to mass per given volume (commonly kg C m^{-2} to a 1 m depth), one must use data on the bulk density of the soil and its rock fragment content. If we are to calculate total organic C storage to 1 m, for each soil horizon we must (1) determine the volume, (2) subtract from the total volume the volume occupied by rock fragments (chemical analyses of C and other properties are only made on the <2 mm soil fraction, as rocks are assumed to make a negligible contribution to these properties), (3) calculate the total mass of <2 mm diameter soil in the horizon, and (4) finally, calculate the mass of C in that layer. This is repeated, and summed, for all soil horizons (or portions of soil horizons) within the upper 100 cm of the soil. Mathematically:

$$\text{mass C/horizon (g m}^{-3}) = [[\text{horizon thickness (cm)} \times \text{horizon area (10,000 cm}^2)] - \text{gravel volume (cm}^3)]$$
$$\times \text{ bulk density (g cm}^{-3}) \times \text{C\%/100}$$

Horizon volume is the product of thickness and chosen reference area. Rock volume is calculated by converting the weight percentage of coarse fragments ($\geq$2 mm) to volume:

$$\text{Volume fraction gravel} = (\text{wt fraction gravel/bulk density gravel})/[(\text{wt fraction gravel/bulk density gravel})$$
$$+ (\text{wt fraction <2 mm/bulk density soil})]$$

It may occur in some problems that percent C (or some other chemical property) is reported but bulk density is not. If one wishes to proceed with a general "back of the envelope calculation," a knowledge of the common ranges in bulk density for soils in that location (and combination of state factors) may be obtained from KSSL databases or other sources. If these are not present, then a second alternative is to conduct calculations for a reasonable range of bulk densities, which allows at least some concept of possible error in the estimates.

---

solubility is related to pH and the presence of other minerals, a series of extractions have been developed for different environments. Briefly, these include:

Bray: P released by an acidic extract, used for soils with slightly acid pH.
Olson: P released by sodium bicarbonate, used for soils with high pH.

NZ (New Zealand P retention): for soils with a high P retention capability, the amount of P adsorbed using this method is the metric of P availability.

Later, the relationship between Bray P and total P for the Memphis soil is examined.

5.2.9 Tier 9. Clay Mineralogy

Quantitative determination of clay mineralogy is somewhat difficult. Several methods of clay mineral identification may be used to better understand the distribution of minerals in this fraction. There follows a very cursory discussion of these methods and their interpretation. The SSL manuals and books on soil mineralogy[5] should be consulted to better understand these methods.

- X-ray diffraction: Horizontally oriented clay mineral samples, when subjected to X-rays, diffract the X-rays in patterns related to the interatomic distances in the crystals of the minerals. X-ray diffraction spectra therefore allow qualitative or semiquantitative assessments of the types of clay minerals in a soil clay fraction.
- Differential thermal analysis (DTA): In DTA, clay samples and inert reference materials are placed in separate platinum pans, and thermocouples measure the temperature differential between the samples as the temperature of the reaction chamber is increased. The temperatures at which exothermic or endothermic reactions occur identify the type of mineral(s) present, and the size of the difference reveals information about the amount of mineral.
- Thermogravimetric analysis (TGA): In TGA, the weight of a sample is measured continuously during heating. Specific minerals undergo reactions resulting in weight loss at defined temperatures, and TGA data provide information on the type and amount of minerals present.
- Total elemental analyses: The clay fraction is completely dissolved in acids and total quantities of major elements are then measured. Ratios of elements (e.g. Si/Al) provide insight into the nature of minerals and their degree of weathering, K amounts are suggestive of illite (mica) amounts, etc. Some data sheets from special studies report the total chemical composition of the whole soil fraction, which is useful in mass balance studies (to be described in Chapter 8) and for general geochemical comparisons.
- Ethylene glycol monoethyl ether (EGME) retention: EGME retention measures the surface area of a clay sample. Surface area per mass is related to the clay mineralogy; for example, smectite has a specific surface of about $10^2 \, m^2 \, g^{-1}$ while kaolinite and mica are around $10^1 \, m^2 \, g^{-1}$.[6] In the EGME method, the weight of EGME adsorbed as a monolayer is proportional to specific surface.

For the Memphis soil, X-ray diffraction indicates that the clay fraction is relatively mixed, with significant smectite present. Additionally, the soil has kaolinite, mica, vermiculite, and quartz. This is all consistent with the CEC/mass of clay discussed earlier.

5.2.10 Tier 10. Sand and Silt Mineralogy

X-ray, DTA, and TGA are occasionally performed on sand and silt fractions as described in the previous section. In addition to these analyses, optical techniques are sometimes performed to determine aspects of sand and silt mineralogy. From optical methods, two data sets are reported:

- Total resistant minerals (TOT RE): This is the fraction of the total mineral fraction made up of quartz, Ti and Zr oxides, etc. This is an indicator of the degree of weathering of the sample.
- Grain count percents: The percentage of given minerals, out of the total grains counted, is reported.

In the case of the Memphis soil, the mineralogy of the silt fraction was analyzed by grain counts. Loess (comprised largely of silt) is commonly dust that is deflated from river channels during winters in glacial periods. Weathering and winnowing of minerals during both river and wind transport create a unique mineralogical signature: quartz is nearly constant with depth (about 65 percent), followed by potassium feldspar, biotite and a suite of other minerals (including zircon). This is indicative of a consistent mineralogical makeup over the depositional time of the loess.

5.3 An Overview of the Memphis Soil History

By becoming familiar with soil data sheets it is possible to use them to develop a biogeochemical perspective of the Memphis soil. To begin, the geographical setting of the soil is examined.

The Memphis area lies to the east of the Mississippi River. During glacial episodes, the Laurentide ice sheet ground and incorporated sediment as it expanded southward from present-day Canada. As melting occurred, the waters carried sediment with it as it moved toward the Gulf of Mexico. During the winters, the stream flow greatly declined, and wind deflation of the Mississippi floodplain spread silt and very fine sand over the downwind areas to the east (Figure 5.2). This loess is dominated by quartz in the sand and silt mineralogy, with other minerals that are more susceptible to chemical weathering (such as potassium feldspar, biotite, etc.). Loess from the last glacial episode in Tennessee has been dated to have begun accumulating between roughly 20 and 25 Ky and continued over time, culminating about 10 Ky ago.[7] This roughly homogeneous blanket of "dust" over a preexisting landscape represents the starting point for the Memphis soil. The Memphis soil is found on nearly 1 million ha, from Kentucky to Louisiana, and is thus a major soil along the Mississippi corridor.

Memphis TN has a mean annual temperature of 17 °C and receives nearly 1400 mm of precipitation annually (Figure 5.3). It has a natural flora dominated by tree cover. Using a Thornthwaite model to calculate monthly evapotranspiration, the calculations show that

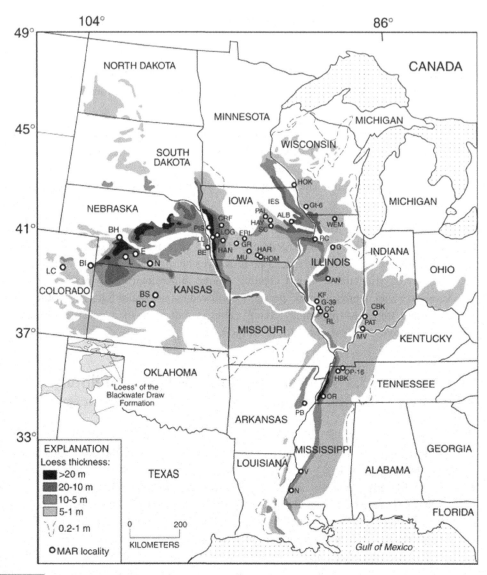

Figure 5.2 Distribution of last glacial loess in the midcontinent of North America. From E. A. Bettis et al., Last glacial loess in the conterminous USA, *Quaternary Science Reviews*, 22: 1907–1946 (2003).

precipitation exceeds evapotranspiration except during the summer and early fall months. The soil has a water-holding capacity of about 500 mm. By calculating, and summing for the year, the monthly precipitation-PET amounts, it is revealed that 511 mm of water remains and moves through the soil each year, enough to fully leach the upper 2 m on an annual basis. This annual flushing of the soil with water – which drives chemical reactions – weathers minerals and potentially removes nutrients.

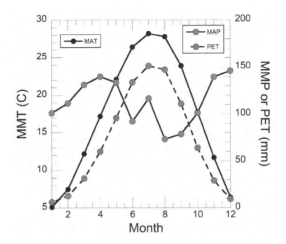

Figure 5.3 The monthly temperature, rainfall, and potential evapotranspiration for Memphis, TN. Data from Wikipedia, and PET calculated from Thornthwaite model.

As will be discussed later, biological soil processes tend to occur more rapidly than most chemical and mineralogical changes. Thus, the accumulation of organic C in the upper horizons of the soil is an important process that occurs over 10^2 to 10^3 y. The profile of the Memphis soil examined here has been farmed, and it will be shown later that cultivation tends to reduce soil C and N by 40 percent or more over several decades. The total mass of C in the soil (using calculations in Box 5.1) is just over 8 kg C m^{-2} in the upper 2 m of the profile. This is a modest quantity (the global soil average is about 10 kg C m^{-2}).[8]

The pH and exchangeable acidity data indicate that there are protons available for chemical weathering. Acids, combined with an adequate flow of water to maintain concentrations below equilibrium conditions, drive rates of chemical weathering.[9] The horizon names (Bt) signify a subsurface accumulation of clay. Clay is present in the regional loess that originally mantled the area but has also been formed and redistributed by soil processes. One can make a first-order assessment of how important each has been. If it is assumed that the lowest Bt horizon is the closest to the original loess composition (we are restricted to this as the C horizon clearly forms from different parent materials), then the entire profile once began with ~21 percent clay. The upper two horizons have a deficit, while the next three have apparent gains. Using the principles in Box 5.1, the upper 2 m has apparently gained ~13 kg of clay above an assumed 21 percent at t = 0 (Figure 5.4a). The apparent gain of clay from chemical weathering is likely reflected in the change in fine to coarse silt. The finer particles (with high surface area to mass ratios) should be more susceptible to chemical weathering, and indeed the low ratios at the surface, which increase with depth, likely reflect this process (Figure 5.4b).

In later chapters (Chapter 8) the interpretation and use of elemental chemistry is explored. This was not performed for the Memphis soil by the NRCS. However, Muhs et al.[10] and Williams et al.[11] examined the total elemental chemistry of a Memphis soil about 40 km from the location of the soil in Table 5.1, which provides an opportunity to bridge the extractive

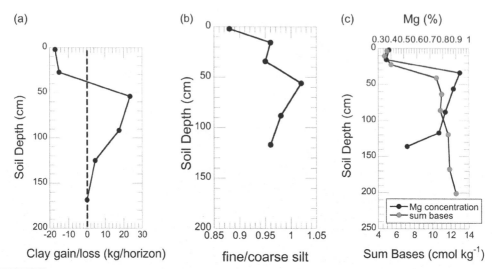

(a) Calculated losses or gains of clay (kg/horizon) vs. depth, (b) ratio of fine to coarse clay with depth, and (c) total Mg in soil vs. sum of bases in the Memphis soil.

chemistry, and what it tells us, to total chemistry. In the Memphis soil, exchangeable bases are low in the upper two horizons and then increase rapidly with depth and reach an approximate constant value (Figure 5.4c). As discussed in Chapter 8, this pattern is characteristic of a downward-moving weathering front through a soil, where dilute waters react with minerals until the waters reach equilibrium and then are largely unreactive as they pass through the remainder of the soil into ground or surface waters. The total chemical data (using Mg as an example) show that this is indeed the case. Thus, for the Memphis soil, the cations adsorbed to the clays at a given depth are indicative of the mineral reservoir at that depth. Some elements that are heavily involved in biogeochemistry (such as Ca and K) tend to have lower depletions in both total and adsorbed phases near the surface but still reflect the effect of the unrelenting processes of weathering driven by rainwater and acids on the soil chemistry. The biocycling of base cations is reflected in the soil pH, which is higher at the surface than in the horizons below. The biocycling of P is probably the most visible due to the relatively low solubility of PO_4^{-3}. As both the Bray 1 (an extract) and total P show, the surface horizon of the Memphis soil is enriched in P relative to the second horizon (due to plant returns of P and possibly fertilization), but both are depleted relative to the underlying soil layers, which reside below the chemical weathering front that is slowly moving downward (Figure 5.5). The rate of this advance is roughly the depth of maximum depletion (~25 cm) divided by the time since loess deposition ceased (roughly 10,000 y), which is thus 2.5 cm Ky^{-1}.

Along with the chemical weathering and accumulation of clay, the colors of the Bt have a distinctive reddish-brown color, which corresponds to the higher concentrations of Fe oxides in these layers (Figure 5.6a). The Fe oxides (likely a mixture of goethite, hematite, and ferrihydrite) now make up about 50 percent of the total Fe in the soil (Figure 5.6b). While some Fe oxide may have been present in the original loess, the correlation with clay

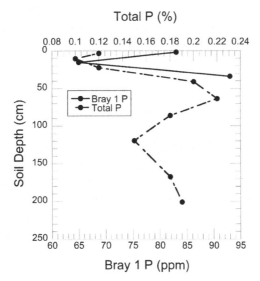

Figure 5.5 Bray 1 extractable P (ppm) and total P (%) vs. depth for the Memphis soil.

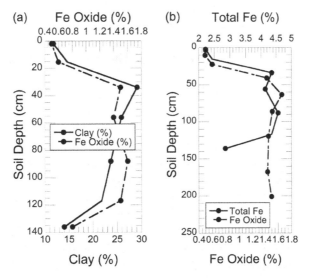

Figure 5.6 (a) Clay % and Fe oxide % vs. depth, and (b) total Fe and Fe oxide vs. depth, for the Memphis soil.

attests to the relatively rapid oxidation of the reduced Fe that is found in the primary minerals.

This cursory overview of the formation of the Memphis soil using NRCS data provides considerable insight into soil biogeochemistry and the processes that have affected it. Without having visited Tennessee, or viewed a soil exposure around Graceland, one can develop a solid understanding of the soil and its history. Most importantly, it illustrated the enormous wealth of information about soil biogeochemistry that is available to users simply through web-based enquiries.

5.4 Summary

This chapter provided an overview of the major types of analyses and data reported on data sheets prepared by the KSSL. These forms contain a wealth of information, obtained at great time and cost, that has not always been easily utilized by students and researchers because of access difficulty. However, the availability of these data on the World Wide Web, due to a long-term effort by the SSL staff, now makes these sheets available to a wide array of potential users. The remaining five chapters in this book now focus on how to understand and examine soils using these data.

5.5 Activities

The data for these activities can be accessed at the NRCS soil characterization data mart: https://ncsslabdatamart.sc.egov.usda.gov/querypage.aspx

5.5.1 Access Soil Characterization Data

The portal has a map of the world with all the soils that are in the NRCS data set. On clicking on any of the points, a pop-up window provides access to soil data, profile descriptions, and other information. At the present time, the soil characterization data are not downloadable in their entirety as in a spreadsheet format. One option is to copy all the data and paste them into Excel as text, creating a workable format. Alternatively, by identifying a soil series name of interest, this can be accessed through the basic query portal on the website.

In this part of the exercise, download the characterization data for a soil of interest. For example, the city and surrounding area of Lahore, Pakistan, resides on a Pleistocene terrace of the Ravi River, a tributary that eventually feeds into the Indus River. The soil series "Lyallpur," Pedon # 84P0572, was sampled by the USDA in 1984. The Ravi River region, and the Indus in general, was the home of one of the earliest major urban and agricultural civilizations on earth, the so-called Indus Civilization (https://en.wikipedia.org/wiki/Indus_Valley_Civilisation).

Plot the clay, silt, organic C, and $CaCO_3$ content vs. depth (of the <2 mm fraction). For the depth of each horizon, many scientists use the midpoint of the horizon as the depth reference. The high silt is derived from the river alluvium, which brings material from the siltstones and other rocks upslope. Lahore is very hot (mean annual temperature = 24.4 °C), and most of the rainfall (~550 mm y^{-1}, from the south Asian monsoon) falls during the hot summer months.

a. Compared with the Memphis soil discussed in the chapter, how does the organic C profile compare? Provide some preliminary hypotheses for any differences.
b. What does the depth profile of clay percentage suggest in terms of chemical weathering and clay transport?
c. What might explain the depth profile of $CaCO_3$ in the soil?
d. By examining the CEC to clay ratio, what type of clay mineralogy might be expected, and how does this compare with the clay abundances determined by X-ray diffraction?

5.5.2 Use of NRCS to Probe Science Questions

One attribute of the website is that it allows the user to probe soil data by depth instantaneously, using the "soil properties by depth" option, once you have accessed some soils. To get experience with this, let's explore the question of how the depth (and amount) of $CaCO_3$ in soil varies with rainfall. To do this, we will loosely use the study design of Jenny and Leonard (1933), who sampled soils along an E to W transect of the Great Plains of the USA. They found that the depth to the $CaCO_3$-enriched layer decreased in a westerly direction as rainfall declined. The advantage of this study design was that much of the region is covered with Holocene loess (constant parent material) and the mean annual temperature is roughly constant (since the latitude is constant). We can rapidly explore this by comparing depth profiles of many soils in eastern Kansas with those of western Kansas. To access the soils in western Kansas, one can use the Advanced Query portal and identify a rectangle covering much of the western part of the state:

✉ **Sign up for E-mail updates on the NCSS Lab Data Mart**

NCSS Soil Characterization Advanced Query Interface

Clear All Search Criteria

Project Information ⓘ

Laboratory Project Name
Project Type   Fiscal Year Country   State or Other Administrative Division   Project Seq #

Submitted Name
Soil Survey Regional (SSR) Office
Submission Date   (Mon/DD/YYYY)   through
Database Source

Site Information ⓘ

User Site ID

Latitude   Longitude
Direction Deg   Min   Sec   Sec   Direction Deg   Min   Sec   Sec
north   +/-   west   +/-
--OR--
Latitude Decimal Degrees   Longitude Decimal Degrees
38.5   +/- 1   -102   +/- 1
Site Area Selection
Country   State or Other Administrative Division   County

Soil Survey Area
Major Land Resource Area
National Park System Land
National Forest System Land

By checking all the soils (this area was chosen to access < the 200 maximum soils that can be graphed at one time), one clicks on the depth report request:

☑ 87P0029	S1986KS093013	Ulysses	Ulysses	87
☑ 88P0864	S1988KS193002	Ulysses	(unnamed)	88
☑ 91P0810	S1991KS081009	Keith	Keith	89
☑ 91P0811	S1991KS081010	Buffalo Park	Ulysses	90
☑ 91P0812	S1991ks081011	Manter	Dalhart	91
☑ 95P0481	S1995KS181002	Ulysses		92
☑ 95P0482	S1995KS181003	Blackwood		93
☑ 95P0486	S1995KS193005	Ulysses		94
☑ 00P0076	S1999KS067002	Richfield	Keith	95
☑ 01N0575	S2000KS181001	Pleasant		96
☑ 01P0087	S2001KS071001	Colby		97
☑ 03N0106	S2002KS199001	Kuma		98
☑ 03N0107	S2002KS199002	Kuma		99
☑ 03N0108	S2002KS199003	Kuma		100
☑ 05N0784	S2005KS193007	Ulysses		101
☑ 07N0502	S2007KS199001	Sweetwater		102
☑ 07N0503	S2007KS199002	Caruso		103
☑ 10N0848	S2010CO009001	Wiley		104
☑ 10N0849	S2010CO009002	Wiley		105
☑ 10N1325	S2010KS075003	Colby		106
☑ 11N0413	S2011CO099001	Nepesta	Nepesta	107
☑ 11N0414	S2011CO099002	Ulysses		108
☑ 11N0415	S2011CO099003	Rocky Ford		109
☑ 11N0416	S2011CO099004	Ulysses		110
☑ 11N0417	S2011CO099005	SND		111
☑ 13N0474	S2012CO099001	Wiley		112
☑ 14N0264	S2013KS203001	Ulysses		113
☑ 17N0455	S2016CO101002	Wilid		114

○ Generate Report   Total: 114

○ Download Data

○ Summary Report

◉ Soil Properties by Depth Report

Usage Note: Use the "Return to Last Data Interface" button when going back one page rather than

Continue	Return to Last Data Interface	Top

Then, choose the data "carbonate, <2 mm" and "overlay pedons," and one then arrives at a result.

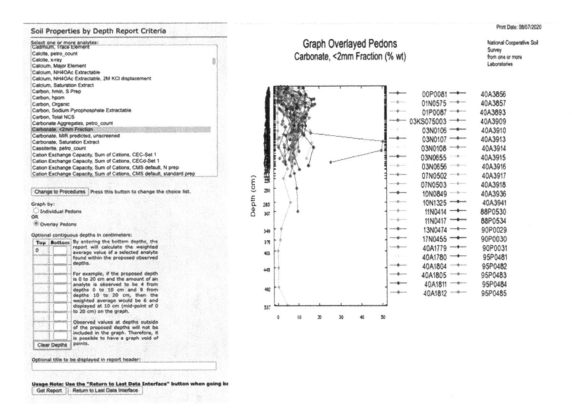

After acquiring the data for the east and western portions of the state, discuss the differences in concentrations, and depth trends, for the two regions (recognizing that a few soils, due to a variety of reasons, fall outside the general trends).

5.5.3 Working with Data

Most soil analyses provide what might be termed "intensive data" in that the value is not dependent on the size of the soil system being studied. Stated differently, most data are based on concentrations, or mass per unit of soil mass. In environmental issues, many times "extensive properties" are required, for example, the mass of a property, such as organic C, per unit volume of soil. Extensive properties allow one to provide quantitative information relevant to many issues, such as the global C cycle, for example.

Here, we will be interested in a soil volume that has a surface area of 1 m^2 to a depth of 1 m. This is a common reference volume for a number of studies or comparisons, such as global C studies. For two soils from opposite sides of the state (Reading, east; Keith, west) calculate:

a. Mass of C to a depth of 1 m in units of kg C m^{-2}.

b. Access Post et al., *Nature*, 298: 156 (1982). What is the global average soil C content of uncultivated and non-wetland soils in their study?

Table 5.1 NRCS data sheet for Memphis soil series

*** Primary Characterization Data***
(Gibson, Tennessee)

Pedon ID: 85TN053001

Sampled as on Sep 29, 1985:
Revised to:

Memphis: Fine-silty, mixed Typic Hapludalts

General information

SSL
- Project CP86TN036 WESTERN AREA
- Site ID S1985TN053001 Lat: 36° 0' 2.00"north Long: 89° 10' 28.01"west MLRA: 134
- Pedon No. 86P0121
- General Methods 1B1A, 2A1, 2B

United States Department of Agriculture
Natural Resources Conservation Service
National Soil Survey Center
Soil Survey Laboratory
Lincoln, Nebraska 68505-3866

Layer	Horizon	Orig Hzn	Depth (Cm)	Field Label 1	Field Label 2	Field Label 3	Field Texture	Lab Texture
86P00770	Ap	Ap	0-17			SIL	SIL	SIL
86P00771	AE	AE	17-38			SIL	SIL	SIL
86P00772	B111	B11	38-70			SICL	SICL	SICL
86P00773	B112	B12	70-113			SICL	SICL	SIL
86P00774	B12	B12	113-137			SIL	SIL	SIL
86P00775	B13	B13	137-200			SIL	SIL	SIL
86P00776	2C	2C	415-438			SIL	SIL	SIL

Pendon Calculations

Calculation Name	Result	Units of Measure
CEC Activity, CEC7/Clay, Weighted Average	0.58	(N,A)
LE, Whole Soil, Summed to 1m	0	cm/m
Clay, total, Weighted Average	28	% wt
Weighted Particles, 0.1-75mm, 75 mm Base	0	% wt
Volume, >2mm, Weighted Average	0	% vol
	28	% wt

ed averages based on control section:38-88 cm

Tiet 1: Particle Size and rock fragments

PSDA & Rock Fragments

				-1-	-2-	-3-	-4-	-5-	-6-	-7-	-8-	-9-	-10-	-11-	-12-	-13-	-14-	-15-	-16-	-17-	-18-
					(------ Total ------)			(- clay -)	CO3	(--- Silt ----)		(------------ Sand ------------)					(-------------- Rock Fragments (mm) --------------)				>2mm
					Clay	Silt	Sand	Fine		Fine	Coarse	VF	F	M	C	VC	(---- Weight ----)			% of <75mm	wt%
Layer	Depth (cm)	Horz	Prep	Lab Texture	< .002	.002-.05	.05-2	< .0002	< .002	.002-.02	.02-.05	.05-.15	.10-.25	.25-.50	.5-1	1-2	2-5	5-20	20-75	-75	whole soil
				3A1a1a	3A1a1a	3A1a1a	3A1a1a	3A1a1a	.002	3A1a1a	3A1a1a	3A1a1a	3A1a1a	3A1a1a	3A1a1a	3A1a1a	3B1	3B1	3B1	3B1	
86P00770	0-17	Ap	S	si	11.8	82.4	5.8	3.6		38.6	43.8	1.8	1.8	1.6	0.4	0.2	tr	—	—	4	tr
86P00771	17-38	AE	S	sil	14.7	83.3	2.0	5.5		40.8	42.5	1.8	0.7	0.5	—	—	—	—	—	1	—
86P00772	38-70	Bt11	S	sicl	29.1	69.5	1.4	16.2		33.8	35.7	1.1	0.2	0.1	—	—	—	—	—	tr	—
86P00773	70-113	Bt12	S	sil	25.9	73.4	0.7	14.1		37.1	36.3	0.7	—	—	—	—	—	—	—	—	—
86P00774	113-137	Bt2	S	sil	23.7	75.7	0.6	13.4		37.4	38.3	0.6	—	—	—	—	—	—	—	tr	—
86P00775	137-200	Bt3	S	sil	21.8	77.0	1.2	11.9		37.8	39.2	1.1	0.1	—	—	—	—	—	—	tr	—
86P00776	415-438	2C	S	sil	14.0	58.8	27.2	8.0		23.3	35.5	0.7	8.7	16.2	1.6	—	—	—	—	27	—

Tier 2: Bulk Density & Moisture

Layer	Depth (cm)	Horz	Prep	-1- (Bulk Density) 33 kPa g cm⁻³ 4A1d	-2- Oven Dry 4A1h	-3- Cole Whole Soil	-4- 6 kPa	-5- 10 kPa	-6- Water Contect 33 kPa act of <2mm 4B1c	-7- 1500 kPa 3C2a1a	-8- 1500 kPa Moist	-9- (Air Dry-Oven Dry) Ratio 3D1	-10- Corrected Soil	-11- WRD Whole Soil cm³ cm⁻³ 4C1	-12- Aggst Stabl 2-0.5mm %	-13- Ratio/Clay CEC7 8D1	-14- 1500 kPa 8D1
86P00770	0-17	Ap	S	1.54	1.55	0.002			22.4	5.1		1.005		0.27		0.69	0.43
86P00771	17-38	AE	S	1.50	1.54	0.009			22.3	5.9		1.005		0.25		0.45	0.40
86P00772	38-70	Bt11	S	1.43	1.60	0.038			28.2	12.9		1.014		0.22		0.55	0.44
86P00773	70-113	Btt2	S	1.45	1.54	0.020			27.7	11.9		1.015		0.23		0.63	0.46
86P00774	113-137	Bt2	S	1.44	1.52	0.018			27.6	11.3		1.014		0.23		0.66	0.48
86P00775	137-200	Bt3	S	1.39	1.48	0.021			28.6	10.4		1.013		0.25		0.67	0.48
86P00776	415-438	2C	S							5.8		1.006				0.68	0.41

Tier 3: Water Content

Layer	Depth (cm)	Horz	Prep	-1- (Atterberg Limits) LL pct <0.4mm 4F1	-2- PI 4F	-3- (Bulk Density) Field 33 kPa g cm⁻³	-4- Recon 33 kPa	-5- Recon Oven Dry	-6- Water Content Field 33 KPa	-7- Recon 33 KPa	-8- 6 KPa	-9- 10 KPa	-10- 33 KPa Sieved Samples	-11- 100 KPa	-12- 200 KPa	-13- 500 KPa % of <2mm
86P00770	0-17	Ap	S	26	4											
86P00773	70-113	Btt2	S	41	18											
86P00775	137-200	Bt3	S	39	16											

Tier 4: Carbon & Extractions

Layer	Depth (cm)	Horz	Prep	-1- (Total) C % of <2mm 6B3a	-2- N	-3- S	-4- Est OC % of <2mm	-5- OC (WB) 6A1c	-6- CN Ratio	-7- (Dith-Cit Ext) Fe 6C2b	-8- Al 6G7a	-9- Mn 6D2a	-10- Al+½Fe	-11- ODOE	-12- (Ammonium Oxalate Extraction) Fe mg kg⁻¹	-13- Al	-14- Si	-15- Mn	-16- (Na Pyro-Phosphate) C % of <2mm	-17- Fe	-18- Al	-19- Mn
86P00770	0-17	Ap	S	0.081				0.83	10	0.5	0.1	0.1										
86P00771	17-38	AE	S	0.045				0.24	5	0.6	0.1	0.1										
86P00772	38-70	Bt11	S	0.052				0.36	7	1.5	0.2	0.1										
86P00773	70-113	Btt2	S					0.25		1.4	0.1	0.1										
86P00774	113-137	Bt2	S					0.14		1.6	0.2	0.1										
86P00775	137-200	Bt3	S					0.20		1.5	0.1	0.1										
86P00776	415-438	2C	S					0.19		0.8	0.1	tr										

Tier 5: CEC & Bases

Layer	Depth (cm)	Horz	Prep	Ca 6N2e	Mg 6O2d	Na 6P2b	K 6Q2b	Sum Bases	Acidity 6H5a	Extr Al 6G9a	KCl Mn	CEC8 Sum Cats 5A3a	CEC7 NH₄ OAC 5A8b	ECEC Bases +Al 5A3b	Al Sat 5G1	Sum 5C3	NH₄OAC 5C1	Exch Na	SAR
				(--- cmol(+)kg⁻¹ ---)							mg kg⁻¹	(--- cmol(+)kg⁻¹ ---)			(--- % Sat ---)	(--- % Base ---)		%	
86P00770	0-17	Ap	S	4.1	0.7	tr	0.4	5.2	5.4	0.1		10.6	8.1	5.3	2	49	64	−	−
86P00771	17-38	AE	S	3.7	0.8	tr	0.4	4.9	4.6	0.1		9.5	6.6	5.0	2	52	74	−	−
86P00772	38-70	Bt11	S	9.5	2.5	tr	1.0	13.0	7.2			20.2	16.0			64	81	−	−
86P00773	70-113	Btt2	S	8.7	2.9	0.1	0.6	12.3	6.5	0.5		18.8	16.4	12.8	4	65	75	1	1
86P00774	113-137	Bt2	S	7.8	3.0	0.1	0.5	11.4	6.8	0.5		18.2	15.6	11.9	4	63	73	1	1
86P00775	137-200	Bt3	S	7.3	2.8	0.1	0.5	10.7	7.2	0.5		17.9	14.6	11.2	4	60	73	1	1
86P00776	415-438	2C	S	4.9	1.9	0.2	0.2	7.2	3.7			10.9	9.5			66	76	1	2

Tier 6: Salt — Water Extracted From Saturated Paste

Layer	Depth (cm)	Horz	Prep	Ca	Mg	Na	K	CO₃	HCO₃	F	Cl	PO₄	Br	OAC	SO₄	NO₂	NO₃	H₂O	Total Salts %	Elec Cond dSm⁻¹	Elec Cond 1:2 dSm⁻¹
				(--- mmol(+) L⁻¹ ---)				(--- mmol(-) L⁻¹ ---)													
86P00770	0-17	Ap	S																		
86P00771	17-38	AE	S																		
86P00772	38-70	Bt11	S																		
86P00773	70-113	Btt2	S																		
86P00774	113-137	Bt2	S																		
86P00775	137-200	Bt3	S																		
86P00776	415-438	2C	S																		

Tier 7: pH & Carbonates

Layer	Depth (cm)	Horz	Prep	KCl	CaCl₂ 0.01M 1:2 4C1a2a	H₂O 1:1 4C1a2a	Sat Paste	Oxid	F	NaF	Carbonate As CaCO₃ <2mm	<20mm	Gypsum As CaSO₄·2H₂O <2mm	<20mm	Resist ohms cm⁻¹
					(---- pH ----)						(--- % ---)		(--- % ---)		
86P00770	0-17	Ap	S		5.5	5.0									
86P00771	17-38	AE	S		5.5	4.9									
86P00772	38-70	Bt11	S		5.6	4.8									
86P00773	70-113	Btt2	S		5.4	4.7									
86P00774	113-137	Bt2	S		5.3	4.6									
86P00775	137-200	Bt3	S		5.2	4.6									
86P00776	415-438	2C	S		5.6	4.6									

Tier 8: Phosphorous

Layer	Depth (cm)	Horz	Prep	Melanic NZ Index %	Acid Oxal	Anion Exch Resin	Availabe Capacity 1	Bray 1 6S3	Bray 2	Resin	Olsen	H₂O	Citric Acid	Mehlich III	KCl Extr NO₃
								(---- mg kg⁻¹ ----)							
86P00770	0-17	Ap	S					83.0							
86P00771	17-38	AE	S					65.0							
86P00772	38-70	Bt11	S					93.0							

Tier 9:

Clay Mineralogy (<.002 mm)

Layer	Depth (cm)	Horz	Fraction	X-Ray 7A21 (Peak size) -1- -2- -3- -4- -5-	Thermal 7A4a -6- -7- -8- -9- %	Elemental 7C3 SO2 Al2O3 Fe2O3 MgO CaO K2O Na2O -10- -11- -12- -13- -14- -15- -16- %	EGME Retn -17- mg g^{-1}	Interpretation -18-
86P00773	70.0-113.0	B112	tcly	MI 3 MT 2 KK 2 VR 1 QZ	KK 3	10.3 1.5		
86P00775	137.0-200.0	B13	tcly	MT 2 KK 2 MI 2 VR 1		10.3 1.4		
86P00776	415.0-438.0	2C	tcly	VR 3 KK 2 MI 2		9.7 0.9		

FRACTION INTERPRETATION:
tcly - Total clay <0.002 mm

MINERAL INTERPRETATION:
KK Kaolinite   MI Mica   MT Montmorillonite   QZ Quartz   VR Vermiculite

RELATIVE PEAK SIZE: 5 Very Large   4 Large   3 Medium   2 Small   1 Very Small   6 No Peaks

---

Tier 10:

Sand -Silt Mineralogy (2 0-0.002 mm)

Layer	Depth (Cm)	Horz	Fraction	Optical Grain Count 7B1a2 (Tof Re) X-Ray peak size -1- -2- -3- -4- -5- -6- -7- -8- -9- Thermal -10- -11- -12- -13- -14- -15- -16- %	EGME Retn -17- mg g^{-1}	Interpretation -18-
86P00770	0.0-17.0	Ap	csi	72 Qz 68 FK 18 BT 4 MS 4 OP 2 PR 1		
				TM1 OW1 RA1 AM1 ZR tr GN tr		
86P00773	70.0-113.0	Bt12	csi	77 QZ 68 FK 18 BT 6 MS 6 OP 1 ZR 1		
				AM1 TM 1 OW 1 RA 1		
86P00775	137.0-200.0	Bt3	csi	72 QZ 66 FK 17 MS 6 BT 5 OP 2 RA 2		
				OW 1 TM 1 ZR 1 HN tr EP tr		
86P00776	415.0-438.0	2C	csi	72 QZ 63 FK 23 OP 4 ZR 3 MS 2 EP 2		
				AM 1 RA 1 BT 1 GN 1		

RACTION INTERPRETATION:
dsi - Coarse Silt 0 02-0.05 mm

MINERAL INTERPRETATION:

AM Amphibole	BT Biotite	EP Epidote	FX Potassium Feldspar	GN Garnet
HN Hornblende	MS Muscovite	OP Opaques	OW Other Weatherable Minerals	PR Pyroxene
QZ Quartz	RA Resistant Aggregates	TM Tourmaline	ZR Zircon	

c. Compare the average C content of cultivated soils with your global average. What might be some of the reasons for the differences?

d. In later chapters, we will examine in more depth how farming reduces soil C. If we assume that the average C content of the cultivated soils in Post et al. is only 60 percent of its original value, how many Gt of C have been lost from cultivated soils due to farming? Is this consistent with recent estimates (based on a web search)?

Time and Soil Processes

"If you free yourself from the conventional reaction to a quantity like a million years, you free yourself from the boundaries of human time. And then, in a way you do not live at all, but in another way you live forever" – unidentified field geologist

John McPhee, *Basin and Range*[1]

Determining the rates of soil biogeochemical processes requires that the time over which the process occurs can be quantified. Biogeochemical processes occur and are measurable over short to very long time spans. While the physical soil mantle may appear to be relatively static, soils are very dynamic and actively functioning three-dimensional bodies that exchange gas, heat, and mass with their surroundings and that are undergoing internal changes as a result. Soil is part of the Earth's surface, and thus an understanding of time on a geological scale is required to interrogate many soil biogeochemical processes.

This chapter briefly reviews some of the concepts and terms of relative and numerical geologic time. It is a language essential to being a natural scientist. The development of the relative and numerical time framework is considered one of the great achievements of science, and this history of science makes as interesting reading as the history of the Earth it describes.[2]

6.1 Soils and Geomorphic Surfaces

The state factor theory states that soil properties are functionally related to the soil's age as well as other important factors:

$$\frac{dS}{dt} = f'(time)_{climate,organisms,topography,lithology,...} \tag{6.1}$$

To determine how a soil property varies with time (and its first derivative, the rate of change) requires knowing how old the soil is. Yet, how is this accomplished? Geological maps do not always provide information on the duration of soil formation because the age of a soil is not always, or even commonly, the same as the age of a geological unit illustrated on a geologic map. The reason for this is that soil age is more appropriately linked to the age of the *geomorphic surface* into which the soil forms (Figure 6.1). A geomorphic surface is defined as "a mappable landscape element formed during a discrete time period."[3] It can be formed by the deposition of sediment – which is called a *constructional surface*. In this case, the surface will be slightly younger than the age of the geological unit that underlies it. Alternatively, a geomorphic surface may be caused by an erosional event that has cut into the underlying rock or sediment – which is called an *erosional surface*. In this case, the

STABLE ------------------------   DYNAMIC------------

Constructional   **Erosional**

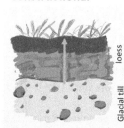

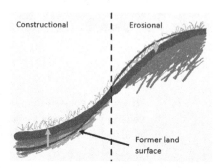

Figure 6.1 A very simple illustration of differing geomorphic surfaces. The so-called "stable" surfaces are those that formed during an event or over time and then have largely (not entirely) stabilized, and soil processes have thus proceeded from the time of stabilization (end of loess deposition or the creation of a river terrace). The "dynamic" surfaces, characteristic of soil-mantled hillslopes, have ongoing erosion on convex portions and ongoing deposition on the concave hillslope positions. Later, we examine the concept of residence time to characterize time of soil formation for dynamic geomorphic surfaces.

surface is younger, or very much younger, than the underlying rock or sediment. In addition, geomorphic surfaces may be constantly evolving, as is the case on hillslopes. As discussed in greater depth in Chapter 9, a parameter called residence time is used to define soil age on constantly evolving surfaces, where the residence time equals the time required for a selected soil thickness to be removed by erosion or buried by incoming sediment:

$$residence\ time(T) = \tau = \frac{soil\ thickness\ (L)}{erosion\ or\ deposition\ rate\ \left(\frac{L}{T}\right)} \qquad (6.2)$$

Soil biogeochemical processes begin once a geomorphic surface is exposed at the Earth–atmosphere boundary or becomes close enough to be affected by the near-surface environment. For example, on dynamic hillslopes (Figure 6.1), the underlying rock may be considerably altered by biogeochemical processes by the time it reaches the land surface in the soil column. In later chapters, the controls on the rate and depth of chemical weathering front propagation through the soil profile will be considered.

6.2 Relative and Numerical Geologic Time

The various divisions of the *relative geologic timescale* are based on the major changes in the history of life on our planet. It is called "relative" in the sense that the major periods and

Table 6.1 An abridged version of the relative geological timescale, with selected biotic and geological events

Era	Period	Epoch	Initial Age (10^6 yr B.P.)	Biotic Events
Cenozoic	Quaternary	Holocene	0.01	extinction of large mammals, spread of modern humans
		Pleistocene	2	early *Homo*
		Pliocene	5	earliest hominid fossils
		Miocene	24	expansion of grasslands
		Oligocene	37	primitive horses and camels
		Eocene	58	early primates
	Tertiary	Paleocene	66	extinction of dinosaurs, expansion of mammals
Mesozoic	Cretaceous		144	
	Jurassic		208	early flowering plants
	Triassic		245	first dinosaurs, early birds and mammals
Paleozoic	Permian		286	coal-forming swamps diminish
	Pennsylvanian		320	coal-forming swamps abundant
	Mississippian		360	first amphibians and reptiles
	Devonian		408	first forests
	Silurian		438	early land plants
	Ordovician		505	invertebrates dominant, first fish
	Cambrian		570	expansive diversification of multicelled life
Precambrian			~3800	origin of life
			~4600	formation of Earth

Modified from Table 8–27 in S. Chernicoff and R. Venkatakrishnan, Geology. An Introduction to Physical Geology, Worth Publishers, New York (1995).

boundaries of the fossil record are empirically observed in the geological record but generally provide no direct information about their numerical age – only the relative order they are found in Earth's history. The *numerical timescale* provides numerical time constraints on the relative timescale, using a variety of "clocks" to provide more quantitative estimates of time. Table 6.1 presents selected components of the relative and numerical timescales. The expanse of Earth's history can be divided into four major segments of time. The Precambrian super-era, an informal unit of geologic time, represents the first four-fifths (~4 billion years) of Earth's history. Life emerged early in this period, altering the atmosphere, the climate system, and biogeochemical cycling. During the long expanse of the Precambrian, the three domains of life[4] that are present today evolved. In particular, the emergence and dominance of bacteria and archaea, life-forms that still dominate the Earth in both numbers and possibly mass,[5] occurred. The remaining 600 million years of Earth's

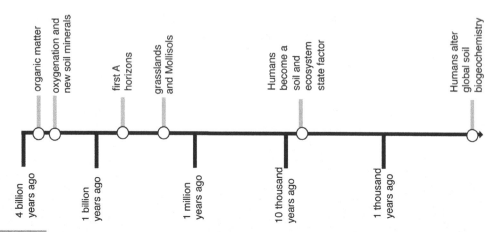

A time line of Earth's history, from the formation of the Earth to the present, identifying some key biogeochemical thresholds in the formation of soils. Concept from the Big History Project (https://school .bighistoryproject.com/pages/console#syllabus)

history is broken into three eras: Paleozoic ("old life"), Mesozoic ("middle life"), and Cenozoic ("recent life"), separated from each other by mass extinctions, whereas the Paleozoic is separated from the Precambrian by the "Cambrian explosion of life," essentially a diversification of the eukaryote (multicellular organism) domain of life.

From the perspective of soil biogeochemistry, the evolution of life marks a fundamental transition, or what is sometimes called a threshold,[6] in the concept of evolution and increasing complexity of the cosmos (Figure 6.2). Other fundamental soil thresholds include the evolution of photosynthesis, which profoundly altered the potential array of chemical processes and minerals possible in soils (Chapter 3). The evolution of land plants in the Silurian (a geological period within the Paleozoic era) marks a fundamental change in the type of biogeochemical processes that soils throughout the remainder of geological time are exposed to. Plants, and their impact on global geochemistry, have been a subject of considerable interest, as they may have driven changes in the rates at which soils consumed CO_2 via chemical weathering.[7] The more recent global expansion of grasslands about 6 to 8 million years ago in the Miocene[8] resulted in a world, and soils, more similar to those of today, where about 25 percent of the land surface comprises Mollisols, or soils usually associated with grassland or steppe.

The Cenozoic era is the most important time period for understanding our present soil distribution. The era is divided into two periods (Tertiary and Quaternary) and seven epochs based on global environmental and geological events (Table 6.1). Probably the most significant event was the onset of global glaciation cycles roughly 2 million years ago, which marks the boundary between the Tertiary and Quaternary periods and the Pliocene and Pleistocene epochs. We are presently in the Holocene epoch of the Quaternary period, one of several brief interglacial interludes in a glacially dominated 2 million-year period. During this time, the northern latitudes of the Earth (as well as montane regions at all latitudes) have been repeatedly scoured by continental glaciers. Other regions have been covered by multiple layers of eolian dust, or fluvial sediments, supplied by glacial meltwater. Finally, erosion and sedimentation cycles in regions far removed from the glacial ice have repeatedly rejuvenated land surfaces in many of the remaining areas of the Earth. These climatically driven events, combined with ongoing tectonic upheavals, leave little of the Earth's surface older than the Quaternary intact. Therefore, the overwhelming percentage of soils are of Pleistocene, or much younger, age. The relatively rare localities on the Earth's surface that are older than the Quaternary offer rare glimpses into the long-term fate of soil biogeochemistry under sometimes unique environmental settings.[9]

However, to fully understand modern soil biogeochemistry, the entrance of humans onto the global landscape must be considered. Beginning with the advent of agriculture, which today directly impacts about 20 percent of the land surface, the Earth's soils began undergoing a domestication process (discussed in detail in the final chapter), which is still evolving. With the development of N and P fertilizers, our impact on soil biogeochemistry began impacting planetary processes, leading into what many consider to be the conclusion of the Holocene, and the beginning of the Anthropocene, sometime in the last century.[10]

While both erosional and depositional events remove soil, burial may preserve the former surficial soil in the geologic record as buried or fossil soils. Many of the Earth's continental sediments are being studied with respect to their fossil soil record. A generic term for fossil soils is *paleosol*.[11]

Determining the relative age of geomorphic surfaces, and hence soil age, has rapidly advanced in the past few decades as a number of geochemical dating techniques have been applied to the problem. With the development of ^{14}C dating in the 1950s the initial advances in numerical age dating of geomorphic surfaces and soils began. Recent years have seen significant advances in soil dating (and their inherent limitations) using ^{14}C of soil organic matter,[12] ^{14}C of pedogenic carbonate,[13] cosmogenic isotope dating,[14] and U/Th dating of pedogenic carbonate.[15]

6.3 Soil Biogeochemical Processes and Time – An Example

There is no perfect or preferred way to quantify rates of soil processes. It depends on the questions one is interested in. One method to determine rates of soil processes is direct observation over time, which provides great insights into short-term questions such as the diurnal or seasonal variations in the rates of specific processes such as carbon dioxide fluxes or water flow. Real-time observations are also becoming much more practical as small, inexpensive sensors become widely available for numerous soil properties (temperature, soil moisture, salinity, CO_2, etc.), data that allow one to monitor heat, water, and gaseous transport processes in great temporal detail.

Yet, the solid phase of soil represents the net biogeochemical changes that have accrued over long time spans, and these in turn provide insights into process rates that are also important for environmentally relevant questions. In particular, *chronosequences*, soils of differing ages with similar sets of remaining soil-forming factors, are highly useful in deciphering rates of processes from features that form over long timescales.

Biogeochemical data from chronosequences are widely available and, with new initiatives such as the Critical Zone Observatory network, will become even more so. Here, two soil processes important on a global scale are examined: (1) a "bio"chemical process: the rates of organic C accumulation, and (2) a "geo"chemical process: the rates of chemical weathering of soil minerals. As discussed in Chapter 2, these are largely silicate minerals. Natural rates of organic C sequestration are useful for determining the rate that soils can remove or otherwise interact with atmospheric CO_2. Rates of mineral weathering are also important for the global C cycle, since the anion usually formed by silicate weathering is bicarbonate (HCO_3^-), which is derived from CO_2. Additionally, rates of mineral weathering are informative for estimating rates of nutrient releases from minerals for plants.

Here, the data from a series of coastal marine terraces on the northern California coast are used[16] to illustrate how chronosequences can be employed to determine rates of these two biogeochemical processes. To begin, the results for these calculations are plotted for the coastal chronosequence in Figure 6.3a and c. The plots show that organic C in the soil increases with age, following a nonlinear, logarithmic pattern. Likewise, cumulative SiO_2 loss increases nonlinearly with age. A function can be fitted through these data to describe how the parameter of interest is correlated with time. However, it is more useful to understand the rates of the processes. To do so, we take the first derivative of the functions describing (a) and (c) and plot them in (b) and (d). The results show that the rates of both C accumulation and chemical weathering of minerals declines with age:

$$\frac{dC}{dt} = \frac{12.60}{t}$$

and

$$\frac{dSiO_2}{dt} = \frac{29.65}{t}$$

These nonlinear trends were expected. First, soil organic C is the balance between plant inputs (which may remain roughly constant with time) and the microbial decomposition of soil C. In many environments, due to the relatively rapid nature of biological processes, a quasi-steady state (i.e. time invariance) is reached in roughly 10^3 y. In Figure 6.3, the organic C continues to increase with time (albeit slowly). A hypothesis for this trend is that in this long time sequence the soils continue to gain clay due to weathering, and this clay enhances the soil's ability to store C at steady state. Thus, to some degree here, organic C storage is also tied and intimately linked to the chemical weathering rates. The rate at which SiO_2 is released also declines with time, largely because the reservoir of weatherable silicate minerals in the soil eventually becomes depleted, and the soil becomes a pool of insoluble quartz and secondary minerals that are in approximate equilibrium with the soil conditions.

The data and calculations in Figure 6.3 can be used to examine the time-dependent relationship between C sequestered by organic matter and by chemical weathering over time (Figure 6.4). Here, for illustration purposes, it is assumed that the mineral being weathered is wollastonite ($CaSiO_3$), which when dissolved releases a mole of Ca for each mole of Si and in turn sequesters 2 moles of HCO_3^-, since Ca^{+2} is a divalent cation:

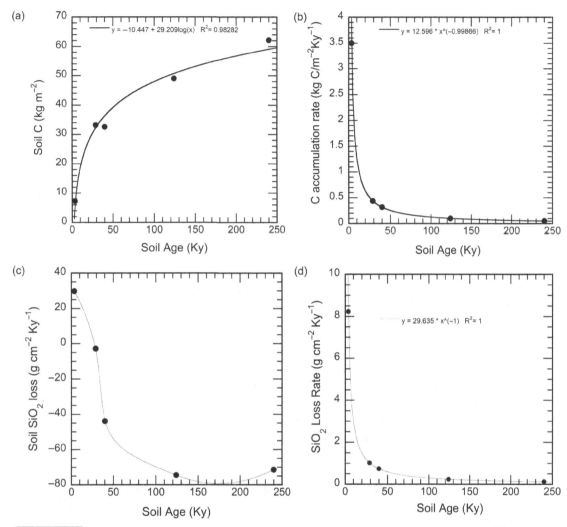

Figure 6.3 Calculations of (a, b) organic C accumulation and (c, d) SiO$_2$ loss using the data of Merritts et al.[14] and mass balance methods discussed in Chapter 8. Data in (c) was multiplied by −1 to allow a logarithmic model to be fitted (as in a), and the derivatives for both (a) and (c) were used to generate rates in (b) and (d).

$$CaSiO_3 + 2CO_2 + 2H_2O = Ca^{+2} + 2HCO_3^- + SiO_2 + H_2O$$

At young soil ages, organic matter C sequestration (due to its fast cycling rates) is large relative to chemical weathering – and cumulative organic C sequestration exceeds cumulative weathering sequestration. However, as time passes and the organic C accumulation trends toward stasis, the remaining pool of minerals continues to undergo weathering, and eventually the C sequestered by weathering exceeds organic matter accumulation by a factor (in this locality) of 5. These results for the humid, cool northern coast of California

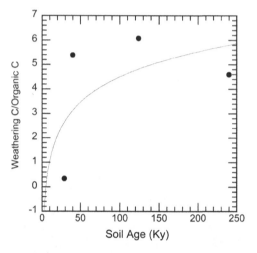

Figure 6.4 A comparison of cumulative C sequestration by chemical weathering vs. soil organic matter accumulation in the coastal terraces of northern California. The logarithmic curve fit reveals that for the very young soils, the rapid rates of organic C accumulation exceed that of weathering, though as soils age, the relative importance of weathering increases.

represent just once slice of the global climate spectrum and cannot be directly extrapolated beyond these conditions. However, growing assemblages of additional chronosequences in a variety of climates and lithologies will help to reveal the underlying time/climate relationships that may be useful for regional or global extrapolations. In Chapters 7 and 8, weathering, the soil C cycle, and the calculations required to probe these processes are examined in more detail.

6.4 An Introduction to the Dating of Soils

Before leaving the discussion of geological time and the importance of soil age, it is important to explore at least one example of dating soil. This topic could be a book unto itself, introducing and reviewing various uses of radioactive isotopes for age control. Here, the use of cosmogenically produced radionuclides to determine soil exposure times is illustrated. At the same time, this helps to understand some of the challenges or complications involved in all dating strategies, none of which are commonly insurmountable.

In Chapter 8, the changes in the chemistry of soils on marine terraces on the coast of California near Santa Cruz will be examined in more detail (Figure 6.5). Marine terraces form along tectonically uplifting coastal margins, for example, the west coast of North America. The uplift, combined with the lowering of sea level by about 100 m during glacial ice ages, allows benches that were cut by waves to be lifted upward and escape erosion once sea levels return to interglacial levels. The uplifted beaches form a terrace. The number of terraces along a coastline is related to the rate and duration of uplift and erosional processes that eventually

Figure 6.5   Aerial photograph of the coastline north of Santa Cruz, CA, with the locations of marine terraces
1 to 4 identified. www.mobileranger.com/santacruz/the-cool-staircase-shaped-hills-north-of-santa-cruz/

remove the terraces. The Santa Cruz coastline, north of the city, exhibits five terraces that all
have some well-preserved segments. It is clear, in a relative sense, that the age of the terraces
(and their soils) becomes greater as one ascends the "staircase," yet determining the absolute
age has been difficult since there is an absence of fossils or other uniquely datable material to
constrain their ages. This is likely a common challenge in many soils.

During the past three decades or so, chemists and physicists have developed a powerful
set of new "clocks" to date soils and rocks: cosmogenically produced radionuclides. Briefly,
cosmic rays from supernovae constantly cascade through the Earth's atmosphere, ultimately becoming largely a rain of neutrons that ultimately impact rocks and minerals (and
soils) at the Earth's surface. Spallation reactions (Figure 6.6) in the mineral quartz (with
both the elements O and Si) result in the formation of radioactive ^{10}Be (half-life 1.4 My)
and ^{26}Al (half-life 717 Ky). The neutrons' penetration power decreases with soil depth and
reaches a low rate of nuclide production at roughly a meter. The rate of cosmic ray impacts
varies with altitude and latitude but is now fairly well constrained, and the rates of nuclide
production in minerals is also constrained. As a result, the concentration of a nuclide, such
as ^{10}Be, vs. soil depth can be calculated as a function of age:[17]

$$N(z,t) = N_o e^{-\lambda t} + \frac{P_o e^{(-\rho z)\Lambda}}{\lambda} \qquad (6.3)$$

Table 6.2 Data to parameterize Eq. (6.1) for the marine terraces at Santa Cruz, CA

Parameter	Value
N_o	1×10^4 atm g^{-1}
λ	4.99×10^{-7} y^{-1}
P_o	4.2 atm g^{-1} y^{-1}
ρ	1.5 g cm^{-3}
Λ	145 g cm^{-2}

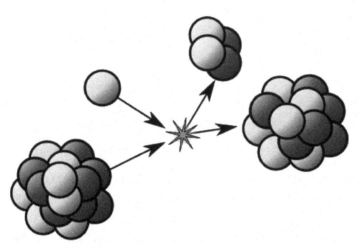

Figure 6.6 An illustration of a single neutron interacting with an atom, forming a He atom (two protons and neutrons) and a radionuclide.

where $N(z,t)$ and N_0 = concentration of nuclide in soil at depth z, time t and time 0 (atoms g^{-1} quartz), P_o = cosmogenic nuclide production rate at the land surface (atoms g^{-1} y^{-1}), ρ = soil bulk density (g cm^{-3}), z = depth (cm), Λ = attenuation depth (g cm^{-2}), λ = decay constant (y^{-1}), and t = time (y).

It should be noted here that most sediments, such as beach sand and gravel, spend some time exposed to the atmosphere (and cosmic rays) prior to being deposited. Thus, many sediments have an initial concentration of radionuclides that must be identified in order to arrive at accurate soil ages.

Perg et al.[18] measured the concentration of ^{10}Be in the five terraces north of Santa Cruz. Table 6.2 lists values for the parameters in Eq. (6.2) taken from both Jungers et al. and Perg et al.[17,18] Perg et al. calculated soil ages from the data using a slightly different method from Eq. (6.2) and additionally assumed that radioactive decay was negligible. Here, decay is included and used in a model to fit curves to compare with the measured data.

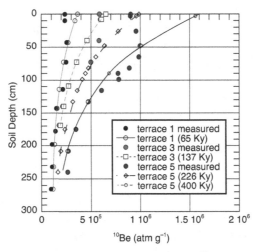

Figure 6.7 The measured (solid dots) and calculated (lines) concentration of ^{10}Be in soils on marine terraces at Santa Cruz, CA. Data from Perg et al. and model parameters used in Eq. (6.1) are found in Table 6.2.

An important observation is that the surface layers of the measured data do not match the model projections (Figure 6.7). This is because the A horizons of the soils are constantly being mixed by organisms and thus do not remain permanently in place as in the model assumption. This process, *bioturbation*, impacts many soil chemical and physical properties and processes. This is likely why Perg et al. calculated soil ages by summing the total amount of ^{10}Be in the soil rather than curve fitting as is done here. Second, the estimated ages provided by Perg et al. produce close curve matches to the data (below the mixed A horizons) for terraces 1 and 3 (2 and 4 were also measured, but for clarity in the figure, only three terraces are examined here). However, the age assigned by Perg et al. (226 Ky) to terrace 5 is not a close match to the data. Here, slightly different production and depth attenuation data are used than those used by Perg. With these data, a much better fit is obtained for a soil age of 400 Ky. While the summation of radionuclides was deemed by Perg et al. to be appropriate, profile matching also is commonly used to interpret measurements. Both methods have errors and uncertainties that extend beyond the introduction here, and this exercise is intended to show some of both the strengths and the uncertainties in any dating method.

6.5 Soils and the Recognition of the Expanse of Geologic Time

The age of soils is increasingly derived from a number of methods that are becoming fairly standard. Yet, recognizing that soil and the Earth record long intervals of time was by no means a simple or easy intellectual step. The quantitative determination of the passage of time is marked through the use of "clocks." The ancient philosophers relied on the celestial clock. Plato wrote: "as a result of . . . .the purpose of god for the birth of time, the sun and moon and the five planets as they are called came into being to define and preserve the measures of time." Of course, today, geological time is recorded through the use of radioactive "clocks" such as the decay of ^{14}C or other isotopes.

The evolutionary biologist Stephen J. Gould wrote of how our concept of an enormous age of the Earth was recognized by the noted Scottish geologist James Hutton.[19] Hutton was in some ways an unusual candidate to promote an unfathomable age for the Earth, a deist who believed in a divine origin of the universe. In accordance with his convictions, Hutton believed that we live in a "world peculiarly adapted to the purpose of man, who inhabits its climates, who measures its extent, and determines its productions at his pleasure." Such a world relies on soil. Hutton also realized that soil formation requires the destruction of rocks: "For this great purpose of the world, the solid structure of this earth must be sacrificed; for, the fertility of our soil depends on the loose and incoherent state of its material."[16] This dilemma – the need for soil production that in turn requires the destruction of the Earth's surface – Gould called "Hutton's paradox of the soil." How can a well-balanced world have both soil and ongoing denudation, which would eventually render the Earth flat and lifeless? The key, Hutton hypothesized, was regenerative forces of uplift and volcanism. The slowness of all these processes, combined with the variety of rocks in every stage of this endless cycle, invoked enormous magnitudes of time.

So, some suggest that soil, and some of its biogeochemical processes, which were understood by Hutton during his 14-year period as a farmer, played a prominent role in one of the great intellectual achievements of all time. This set of historical events, and the geologic timescale (Table 6.1 and Figure 6.2) itself, may now seem inherently obvious from our vantage point. But step back to that moment to view a rock outcrop with John Playfair, Hutton's friend and colleague, in order to see what a first viewing of a geological exposure revealed in light of this revolutionary view of the Earth:

> On us who saw these phenomena for the first time, the impression made will not easily be forgotten ... We felt ourselves necessarily carried back to the time when the schistus on which we stood was yet at the bottom of the sea, and when the sandstone before us was only beginning to be deposited ... Revolutions still more remote appeared in the distance of this extraordinary perspective. The mind seemed to grow giddy by looking so far into the abyss of time.[18]

Soils are also a part of our surroundings that, with appropriate concepts, can be viewed with excitement and a fresh perspective that is also ultimately intellectually transformative.

6.6 Summary

No "Factor of Soil Formation" is more important than any other. The focus here on time is due to its complexity and to conceptual and numerical advances over the decades that have given Earth scientists the amazing ability to use radioactive and stable isotopes to date, with increasing precision, many landscape features. The additional utility of time, chronosequences, and the corresponding mathematical functions is that they provide field-based estimates of the rates of processes: organic matter accumulation rates, rates of chemical losses, etc. This is powerful information, and the ability to determine this has occurred only in the lifetime, and careers, of many scientists working today.

6.7 Activities

6.7.1 Identifying a Chronosequence

Soils of differing ages, in relatively close proximity to each other (and on the same parent material, etc.), are found in a restricted set of geological conditions: river terraces, marine terraces, glacial till/moraines, dune sequences, and a few other types of situation. In this section, we examine terraces of the Merced River of California, which like all Sierran rivers, downcuts into the upper reaches of its channel over geological time due to uplift of the Sierra (its watershed) and slow subsidence in the San Joaquin Valley (its approximate base level). This chronosequence has been the focus of a significant number of interrelated projects over the past 30 years.

The following map below shows the Sierra Nevada to the east and the alluvial deposits and terraces to the west. The dots show soil sampling sites that form the basis for the research projects. A key tool for Earth scientists is Google Earth. Cross sections with coarse elevation profiles can be generated to observe topographic changes with location, showing how the oldest terraces are now the highest landforms.

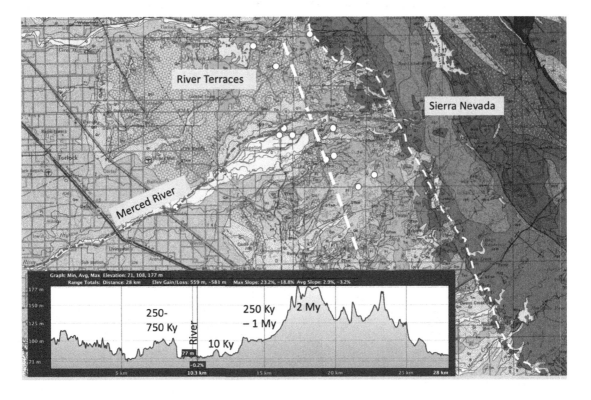

6.7.2 Soil Data

In a few places around the world, landforms, and the soils on them, persist through long periods of time. One of these areas is the Merced River terraces of central California. These terraces formed due to a combination of uplift of the Sierra Nevada (the older terraces) and glacial outwash episodes (the later terraces) over the past 2+ million years. The parent material is granitic alluvium; the climate is semiarid (mean annual precipitation (MAP) = 350 mm y^{-1}; summer dry, winter moist) and warm. Here, selected soils from this study are compiled.

a. Clay
b. Fe_{ox}
c. pH
d. CEC

Calculate the mass of clay and Fe (dithionite) in the upper 2 m. Calculate a depth-weighted pH and CEC to 1 m. Plot each of these and provide a brief explanation of the processes that cause these trends.

6.7.3 Rates

Chronosequences provide natural experiments to understand the rates of processes without resorting to experiments. For these soils, what is the best fit model of clay content vs. time? If it is nonlinear, what is the nature of its change with time, and why would one expect it to behave this way? Calculate the rate of clay accumulation vs. time.

7 The Soil Carbon Cycle

I am fire and air; my other elements I leave for the baser forms of life.

William Shakespeare, *Antony and Cleopatra*

7.1 Introduction

Shortly before his death, Charles Darwin published his final book.[1] Most people today think of it as his "worm book." But the book has a second, and commonly overlooked, subject that is clearly identified in the title: "The Formation of *Vegetable Mould* (italics added here), Through the Action of Worms, with Observations of their Habits." If we were to translate the quaint, nineteenth-century naturalist terminology into modern scientific vernacular, the title would read: "The Formation of *Soil Organic Matter* . . . ." Darwin appeared concerned that in relation to his earlier work on other more monumental topics of scientific concern, the book would appear an aberration, for he wrote:

> The subject may appear an insignificant one, but we shall see that the maxim 'de minimis lex non curat' (the law is not concerned with trifles) does not apply to science.

While Darwin may have intended this statement to apply to the worms, it is equally, if not more, apropos to the subject of soil organic matter in the light of our growing understanding of the role of soil organic matter in the global C cycle, the global N cycle, and nutrient transfers of many kinds. There are many dimensions to the study of soil organic matter: (1) its chemical composition and structure, (2) the molecular pathways by which plant material is converted to these compounds, (3) the patterns of soil organic carbon (SOC) storage on the landscape (both horizontally and vertically), and (4) the rates of the regional- to global-scale processes that control soil organic matter (SOM) and determine its impact on the atmosphere.

Much research in soil microbiology and chemistry has been devoted to topics 1 and 2. It would be fair to state that there is yet much to learn about the molecular-scale processes and transformations of plant and animal matter as it undergoes its pathways of decomposition. The comments Oades wrote several decades ago are still apropos: "Since about 1960 there has been a steady increase in information on humus chemistry, but only an optimist would claim any breakthrough."[2] A conceptual illustration of the flow of plant material from the original plant to its ultimate decomposition product CO_2 is shown in Figure 7.1. The model emphasizes the concept that SOM decomposition is

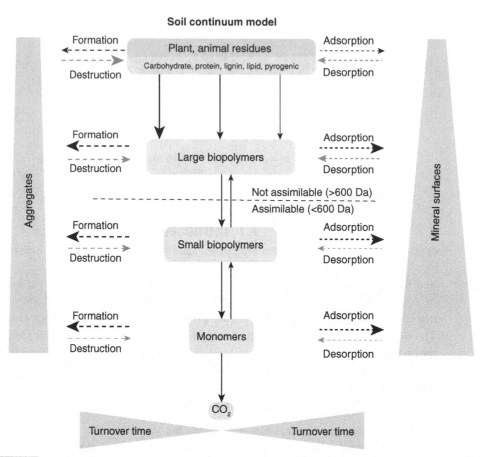

Figure 7.1 A conceptual diagram illustrating the pathway of C from the living plant or animal source, through various interactions with soil microorganisms and the physical environment, to its ultimate conversion back to CO_2 and the atmosphere. There is also a physical transport component that is embedded in this conceptual model. Interactions with the environment and the rate of biological processes dictate the average time (turnover time) that is required before the solid-phase C is respired as CO_2. From J. Lehmann and M. Kleber, The contentious nature of soil organic matter, *Nature*, 528: 60–68 (2015).

a continuum, where free particles of varying molecular sizes and compositions are continuously interacting with the soil physical system and can become embedded in aggregates or adsorbed to mineral surfaces. These temporary interactions with the soil protect the organic compounds from enzymatic decomposition. As discussed later, while individual compounds may rapidly move through this continuum, there is strong evidence that the average pool of organic matter may take a considerable period of time as it passes from the beginning to the end of the pathway.

Briefly, there are significant challenges to understanding the molecular origin and composition of SOC in any given soil, since they contain multitudes of compounds formed over enormous spans of time. In this book, the focus is largely centered on topics 3 and 4: the patterns of SOC storage on the landscape (both horizontally and vertically) and the rates of the regional- to global-scale processes that control SOM and determine its impact on the atmosphere. Rather than focusing on the type and origin of SOC (important questions in their own right), the focus is on how bulk SOM cycles and the resulting macroscopic spatial distribution. Understanding these issues is central to addressing an array of important biogeochemical problems.

7.2 Chemical Composition of Plants

In Chapter 2, the total chemical composition of plants was examined. Plants mimic the overall chemical composition of waters, with the exception of large relative increases in the concentrations of C and N. The large proportion of C in plants is due to the reduction of CO_2 during photosynthesis and the creation of energy-rich C-C bonds in organic compounds. Like rocks, plants are made up of various combinations of elements, but given the dominance of C, H, and O in plants, the compounds are mainly organic rather than mineral. These compounds have differing bond strengths and nutrient contents, which makes their susceptibility to decomposition by microorganisms somewhat variable.

- *Cellulose*: Cellulose is a straight-chain polymer of D-glucose units, thousands of glucose units in length, that is a fundamental unit of cell wall structure and stability. Cellulose content commonly increases with plant age and is particularly high in wood and straw. A wide range of microorganisms contain enzymes (cellulases) to break down cellulose.
- *Hemicellulose*: Hemicellulose consists of polymers of sugars and uronic acid. Pure hemicellulose is easy for many microorganisms to decompose, but in nature hemicellulose may be complexed with compounds and more difficult to decompose.
- *Lignin*: Lignin is a cell wall structural component of plants whose basic building block is a hydroxylated aromatic benzene ring – 500 to 600 such rings are linked in lignin. The chemical composition and amount of lignin vary from plant to plant. Lignin concentration increases with age, and the greatest concentrations of lignin occur in woody plants. Lignin decomposition is primarily by fungi, which are eukaryotic organisms.
- *Protein*: Proteins are composed of repeating units of amino acids (N- and S-bearing organic compounds). The number and combination of amino acids present define individual proteins. A wide range of microorganisms produce enzymes capable of hydrolyzing proteins.
- *Lipids*: Lipids are complex compounds formed from an array of fatty acids. Most lipids are relatively easily degraded by many microorganisms.
- *Minerals and inorganic elements*: While not the subject of as much interest as the organic compounds, most plants contain varying amounts of minerals, and some plants may produce substantial quantities. Members of the grass family *Poacea* produce up to >5 percent (by weight) opal in their leaf tissue.[3] Some plants, such as the hackberry tree of the genus *Celtis*, produce an endocarp in their fruit that is almost pure aragonite ($CaCO_3$) with smaller amounts of opal.[4] While this array of minerals has a small role in

the importance of C cycling, it may have an impact in nutrient cycling, pH control of the soil, and its overall mineralogical composition (e.g. some grassland soils are 3 percent (by weight) opal phytoliths from grass).[5] In addition, inorganic elements are essential plant nutrients involved in a series of biochemical processes.

7.2.1 Decomposition of Plant Compounds

In general, the susceptibility of plants to degradation can vary greatly between species. At a given site, one might find a substantial O horizon beneath one species and almost none under another. Given equal inputs of litter, the differences are due to the palatability of the litter to microorganisms. As a general framework for considering the relative decomposability of litter, the following characteristics of plants should be considered.

- *Lignin content*: As discussed earlier, lignin is a complex molecule that is decomposable by a small subset of the total microbial population in soils. The lignin content of plant litter (or the ratio of lignin to some other plant component) is used in several models to more accurately assess the rate at which the material will decompose.[6]
- C/N ratio: Nitrogen is commonly a limiting element in ecosystems – for both plants and microbes – and the C/N ratio of incoming plant litter will help determine the rate at which it will be decomposed. Plants typically have very high C/N ratios (40:1 or more) while soil microorganisms must achieve a much lower ratio due to the high concentration of proteins in their structure (soil microorganisms can have C/N ratios between 3.5:1 (bacteria) and 15:1 (fungi).
- *Other factors*: Additional plant-specific factors may influence the rate of the decomposition of the material. A high wax content, the presence of specific organic compounds, etc. may all make a specific plant material less susceptible to decomposition.

Next, we consider the characteristics of SOM, the end product of plant material once it enters the soil.

7.3 Processes and Factors Controlling Soil Organic Matter

Once plant material is added to a soil (through one of several processes described later), it becomes part of *soil organic matter*, which is defined as the mixture of plant parts and material that has been altered to the degree that it no longer contains cellular organization.[7] The nonrecognizable material is commonly called *humus*. Because the term humus has had a number of historical meanings that have raised some criticism of its use,[8] the term soil organic matter (SOM) is used here, or in terms of soil organic C, SOC.

The amount of SOM or SOC in a soil, at any given time, reflects the long-term balance between inputs and losses (Figure 7.2). The rates of these processes are controlled by the factors of soil formation. The key to developing an SOC model is to identify the relevant

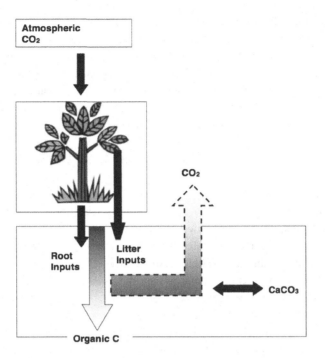

Figure 7.2   Simple schematic of C flow through the soil system.

"pools" and "fluxes" that must ultimately be represented mathematically, which is done in the following sections.

7.3.1 Soil Organic Carbon Pools

Soil organic matter can be thought of in terms of "pools" – i.e. how much SOC is found in a given volume of soil. In this chapter, two approaches will be taken to examine SOM pools: (1) first by considering soils to be a well-mixed box, and (2) by viewing soil as having vertical variations. Total organic C is considered rather than individual components or pools of differing cycling rates. As discussed in Chapter 5, total organic C is now commonly measured by elemental analysis (combustion) following the removal of any carbonate.[9]

7.3.2 Organic Matter Additions

This chapter focuses specifically on C, the major element in SOM. The same approach is generally applicable for N, the second important element in SOM. However, different and additional processes are involved in N cycling. A separate discussion of similar detail for N is not included here. The interested reader should consult papers on soil N cycling[10] that follow the general framework introduced in this chapter.

Carbon (and organic matter in general) is added to soil via (1) surface litter (leaves, branches, crop residue) and (2) roots (root death or exudation of organic compounds from living roots). However, not all of this material may actually find its way into the soil. Much of the aboveground plant material is decomposed by microorganisms to CO_2 before it can be incorporated into the soil. Two major mechanisms to bring SOC into the soil are (1) downward water movement of dissolved or suspended organic particles and (2) direct biological incorporation of surface litter into the soil. Charles Darwin demonstrated the ability of certain earthworms to drag surface litter into their burrows. On a different scale, plowing by farmers is a different way to efficiently incorporate organic substances into soil.

7.3.3 Organic Matter Losses

Most C losses from soils occur as: (1) CO_2 produced by biological decomposition of organic matter, (2) dissolved organic C in water leaching through the soil, and (3) erosion. In most soils, the decomposition and loss as CO_2 is by far the most important means of removing C. In certain climates, however, such as the western and northern regions of Scotland, for example, prodigious amounts of C are lost in a dissolved form – as the chocolate-brown rivers, streams, and springs attest. Finally, the sloping landscapes of the world are constantly undergoing erosive soil loss,[11] and the sediment and C removed by this process are either stored in nearby basins or transported toward the seas.

The CO_2 produced in soils diffuses to the atmosphere, driven by a concentration gradient. The rate of CO_2 evolution from soils, termed soil respiration, can be measured using gas analyzers or other approaches. Total soil respiration consists of CO_2 released by SOM decomposition as well as CO_2 produced by the respiration of living roots. The proportion of the total soil respiration due to decomposition is poorly known, although various approaches suggest it is 50 percent or more of the total respiration.[12]

7.3.4 Soil Organic Matter Transfers within the Soil

The types and magnitudes of processes that move organic matter through the soil are not well examined in most locations. In ecosystems with large earthworm populations, Darwin[13] provided quantitative estimates of net downward transport of stones by their burial by worm casts. Recently, geomorphologists have recognized that biological processes like gopher activity, tree throw, etc. effectively turbate the soil and move it, in conjunction with gravity, downslope. These processes must also move organic matter through the soil.

There are two processes that transport material downward: (1) advection and (2) diffusion. Advection is the transport of SOC by bulk motion, whereas diffusion is the movement of SOC by random motion along a concentration gradient. Both processes probably occur in soil. Advection can occur when organic C dissolves in downward-moving water, which certainly is important in forested ecosystems. On the other hand, diffusion is important in soils slowly mixed by animals, insects, and other biophysical processes.

The rate of C and other elemental movement downward through the soil is beginning to be studied more broadly. The major advancement has been the use of short-lived isotopes made both naturally and by human activity to observe and model the movement of particles. One important method is the use of natural and human-made radioactive isotopes such as

^{210}Pb (half-life $= 22$ y) or ^{7}Be (half-life $= 53$ d), which are both useful to trace and quantify organic matter and particle movement downward. Kaste et al.,[14] in grasslands of coastal central California, calculated a D value of 0.10 cm^2 y. The random biological mixing of soil is similar to diffusion, while the direct movement of particles downward by moving water or due to direct transfer by organisms is an advective process. These have different mathematical forms:

$$\text{Diffusion transport} = -D\frac{d^2C}{dz^2} \tag{7.1}$$

$$\text{Advective transport} = -v\frac{dC}{dz} \tag{7.2}$$

Where D has units of (commonly) cm^{2}y^{-1} and v has units of cm y^{-1}.

Net diffusion in a given layer (Eq. (7.1)) is driven by the change in the concentration gradient with depth, while advective diffusion (Eq. (7.2)) is driven by the concentration gradient. The advective velocity is straightforward to interpret. For diffusion, a characteristic travel time can be estimated by:

$$time\ to\ travel\ distance\ L = L^2 D^{-1} \tag{7.3}$$

Box 7.1	Derivation of C Model

In the development of this model, it is assumed that soil C inputs and losses are constant with time and that, at time $= 0$, the C content is

$$C(t_o) = 0 \tag{7.4}$$

i.e. no C is present in the parent material.
From Eq. (7.10) in the text, the change in C in the soil with time is

$$\frac{dC}{dt} = I - kC \tag{7.5}$$

By separating the variables, the equation can be integrated:

$$\int \frac{dC}{I - kC} = \int dt \tag{7.6}$$

which gives the solution

$$ln|I - kC| = -k(t + K) \tag{7.7a}$$

which rearranged is

$$I - kC(t) = e^{-k(t+K)} = Ke^{-kt} \tag{7.7b}$$

where K $=$ a constant of integration.

Since at t $= 0$, C $= 0$, then
K $=$ I, and substituting into Eq. (7.7b) gives

$$C(t) = \frac{1}{k}(I - Ie^{-kt}) \tag{7.8}$$

which, for soils examined by Kaste et al. of $\sim$ 50 cm, is about 1 to 2 mm per century. Recently, Jagercikova et al.[15] reviewed the literature on short-lived Cs isotope studies and developed a diffusion/advection model to interpret the data to derive rates of transport. They found that D values ranged from 0.02 to 4.44 cm^2y^{-1} (median of 0.64) and advection velocities ranged from 0 to 0.74 cm y^{-1} (median of 0.1 cm y^{-1}). The overall penetration velocities of the Cs isotopes ranged from 0.05 to 0.76 cm y^{-1} (median of 0.28), representing the effect of both diffusion and advection.

7.3.5 Soil Organic Matter Transformations

The conceptual model of the pathways that plant material follows as it breaks down in soils was illustrated in Figure 7.1. A common simplification of this complex set of processes is to assume that the decomposition of organic compounds, and the resulting production of CO_2 and various humic substances, is a first-order decay reaction.[16] This means that the rate of CO_2 production in soil is dependent on the amount of organic C present and that the rate at which a given mass of it decays is determined by a "decay constant." In mathematical terms:

$$CO_2 \text{ production rate} = kC \tag{7.9}$$

where $k = T^{-1}$ and C = mass/area.

In the relatively simple models considered here, the rate at which a pool of organic matter decays is considered constant. However, this is unlikely to occur in nature, since the rate of microbial activity increases systematically with temperature and likely changes with depth due to temperature changes, quality of organic matter changes, and other physical attributes. More mathematically flexible approaches (e.g. numerical models) are needed to explore these more complex cases.

7.4 One-Box Model of Soil Organic Matter Cycling

The processes that drive the cycling of soil carbon can be captured in a simplified mathematical form. The approach is to write a mass balance equation, one that keeps track of inputs and losses over time. An assumption about the amount of C in the soil at t = 0 is required. From this, the change in the amount of C in a soil box (Figure 7.2) with time ($\frac{dC}{dt}$) is the difference between inputs from plants (I) and losses as CO_2 (L, or in the case of a first-order decay approach, kC, where k = fraction of C lost as CO_2 each year (yr^{-1}) and C = mass of C per soil volume (g cm^{-3})):

$$\frac{dC}{dt} = I - kC \tag{7.10}$$

Many real processes are lumped into the two C transport terms. For example, inputs could vary from leaf fall to root inputs, dissolved organic matter input, etc. Additionally, the soil is viewed as a box with no vertical variations. Some additional assumptions in this simple model include: (1) inputs of C to the soil are constant with time and (2) the decomposition constant is constant with time.

The solution to Eq. (7.10) describes soil C as a function of time. By integrating Eq. (7.10) (see Box 7.1 for explanation of this step):

$$C(t) = \frac{1}{k}(I - Ie^{-kt}) \tag{7.11}$$

In this equation, C(t) = mass of C in soil at time t (g cm^{-3}), t = time (yr), and the rest of the terms have been described earlier. Using Eq. (7.11), if the rates of C input to a soil and its rate of decomposition are known through field observations or from lab studies, one can estimate how a soil has gained C over time. There are various ways to measure these parameters, which are briefly described in Box 7.2.

An important version of Eq. (7.10) is when soil is at steady state:

$$\frac{dC}{dt} = 0 = I - kC \tag{7.12}$$

which when rearranged gives:

$$C = I/k \tag{7.13a}$$

$$\tau = 1/k \tag{7.13b}$$

where τ = the soil C residence time, or the time required for inputs or outputs to replace all C in the soil pool. This is an important relationship, in that if one has data on the total C stored in a soil, and an estimate of either inputs or soil C losses (which are equal at steady state), the decomposition rate constant can be calculated. This is used in the next section to show how soils respond to changes in climate.

7.4.1 Box Model Calculation of Variations in Soil Organic C with Climate

The effect of climate on the amount and rate of C cycling is a dominant and timely question in soil biogeochemistry.[17] Equations (7.11) and (7.13a, b) allow one to learn about SOC dynamics if data on total SOC and estimates of inputs or losses are available. The study of soil and ecosystem C cycling is now a massive global effort, and soil C storage data are available, for example, through the Harmonized World Soil Database.[18] Compilations of soil respiration rates have been made over the past few decades and are part of a global set of observational sites devoted to obtaining net ecosystem C fluxes.[19] Here, to illustrate the use of the models introduced in this chapter, data compiled by Sanderman et al.[20] are used. In that study, the authors estimate rates of soil respiration (t C ha^{-1}y^{-1}) from flux tower data (FLUXNET), arriving at the soil component of the total ecosystem respiration. The location of the sites was used, along with a global gridded data set of soil C, to estimate the total soil C storage (kg C m^{-2} to 1 m). As the authors noted, two sites were excluded due to intensive human management (only natural sites are explored here), and two other sites were excluded due to their wet or boggy conditions (only well-drained sites are intended to be compared to examine the impact of climate).

Measuring C Inputs into Soils

- *Litter traps*: Litter traps are trays, baskets, etc. of known surface area that are placed on the land surface in shrub or forest lands to capture leaves, stems, etc. that fall onto the soil surface. The traps are monitored over a period of time to establish the long-term rate of organic matter added to the soil surface. As discussed earlier, this knowledge does not always give a good perspective of rates of soil C input since much of the litter is decomposed to CO_2 before it enters the soil. Of course, it would be best to have complementary knowledge of root death and entry into the soil C cycle, but such measurements are very difficult and expensive to make. Many soil C modelers, as a first approximation, may assume that surface litter addition rates roughly equal the C that actually makes its way into the soil via roots and surface litter combined.
- *Grass clipping:* In annual and perennial grasslands, it is relatively easy to measure the total amount of grass that, at the end of the season, dies and is added to the soil surface. A given area of the land surface (a square meter, for example), is selected, clipped, and the clippings are dried and weighed. Again, this does not give an estimate of what actually enters the soil, which in grasslands is especially dominated by root inputs. One simple approximation is that the grass ("aboveground inputs") equals or is some fraction of the root inputs ("belowground").

These two simple and inexpensive means of measuring C inputs are not exhaustive but are used in a "back of the envelope" level of precision, one that is in many cases very important to have.

Measuring C Losses from Soils

- *CO_2 losses*: If it is assumed that most C is lost from soil as CO_2, then measuring C losses becomes easier. Carbon dioxide produced by microorganisms builds up in the soil and diffuses to the overlying atmosphere by the process of diffusion. One can measure the rate of CO_2 loss from a soil (on an area basis) by installing respiration chambers over the soil surface and then measuring the rate at which CO_2 builds up in the chamber.

While the rate at which CO_2 is released from the soil (soil respiration rate) is possible to measure, it is difficult to interpret. Not only does organic matter breakdown produce CO_2, but living roots of plants respire CO_2 as well. In most places, the proportion of total soil respiration due to organic matter breakdown is poorly known. Many people assume that organic matter breakdown makes up between 50 and 70 percent of the total soil respiration, although that value certainly varies with the time of year (as plants become dormant, they respire at a much lower rate).

First, it is informative to investigate how the soil C at these sites varies with the two key climate parameters: MAT and mean annual precipitation. For the FLUXNET sites examined, mean annual precipitation was modest to high at all sites and was likely not a limiting factor to plant production and soil C inputs. In contrast, there was a 30 K range in MAT, and soil C clearly declined with increasing MAT (Figure 7.3a). From Eq. (7.12), it is known that soil C is the balance between inputs and losses. The rates of soil respiration (CO_2 estimated to be produced by microbial decomposition of SOC) increased with increasing MAT (Figure 7.3b). This inverse relationship leads (Eq. (7.13a, b)) to the result that soil C residence time declines with MAT and k increases with MAT (Figure 7.3c, d) due to the effect of temperature on rates of microbial degradation and the use of soil C as an energy source.

By using the parameters derived previously, Eq. (7.11) can be employed to consider how soil C accumulates with time under differing climate conditions. To illustrate, the data from the coldest and warmest sites in Table 7.1 are used as climate endmembers. The coldest site with the lowest k value requires nearly 1000 y to reach steady state, whereas the warm site achieves this in about 100 y. Many researchers have recognized that one-pool models are approximations of C cycling that underestimate the response of the faster-cycling C in the soil. Nonetheless, these projections agree to a first order with observations of soil C accumulation along soil chronosequences.[21]

7.5 Soil Carbon Distribution with Depth

Treating soil C (or any other parameter) in a one-box model is a useful way to explore how the property is controlled by environmental parameters (particularly climate) and to scale data up to regional or global perspectives.[22] Yet, when examining a profile in the field, or closely examining laboratory data, the clear pattern that emerges is that SOC declines with depth. What drives this pattern, and can these processes further inform and improve our knowledge of soil C cycling? First, just from examining soil profile C data, it is clear that one depth pattern is not universal (Figure 7.4). Soils in deserts, with irregular root distribution with depth, and low and spatially erratic rates of surface inputs, tend to have irregular C changes with depth (Figure 7.4b). In contrast, soils that form in the humid, deciduous to coniferous forests of the more northern latitudes have a thick layer of leaf litter (O horizons) that overlies a brightly leached E horizon. The release of organic molecules in the O horizon, over time, strips away Fe and Al oxides, which precipitate at greater depths once the solution becomes saturated with respect to metal constituents (Figure 7.4a). While there has been research on these later soils and the combination of hydrology and chemistry that forms these patterns, we here focus on probably the most universal SOC pattern in soils, a pattern where C declines nonlinearly with depth (Figure 7.5c).

While this nonlinear (commonly logarithmic) pattern with depth bears some similarity to expected declines in root inputs with depth, the pattern is likely due to much more complex processes. First, soils constantly undergo slow mixing by biological processes. Earthworms, gophers, ants, and other organisms loft soil toward the surface and slowly bury the land surface. In addition, this process leads to mixing of most near-surface layers. This biotic movement mimics the process of diffusion, where molecules or atoms are randomly mixed but have a net movement in the direction of a concentration gradient.

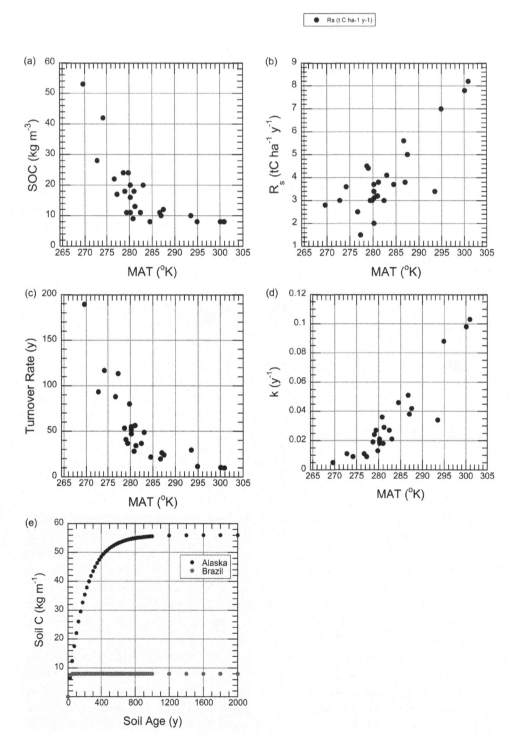

(a) SOC content vs. MAT, (b) soil respiration vs. MAT, (c and d) soil C turnover time and decomposition rate (k) vs. MAT, and (e) model estimates of soil C accumulation over time for the coldest and warmest sites, assuming no C at t = 0. Data from Table 7.1.

Table 7.1 Climate data and estimated soil C and soil respiration (from SOC decomposition) for FLUXNET sites

Site	MAT (K)	MAP (mm)	OC (kg m^{-2})	Rh (tC ha^{-1} y^{-1})
North BOREAS	269.6	420	53	2.8
Prince Albert BOREAS	272.8	368	28	3
Flakaliden, Sweden	274.2	567	42	3.6
Hyytiala, Finland	276.7	640	22	2.5
Renon-Ritten, Italy	277.3	1010	17	1.5
Norunda, Sweden	278.7	527	24	4.5
Bayreuth, Germany	279	885	18	4.4
Takayama, Japan	279.4	1732	11	3
Solling FL, Germany	279.8	1045	24	3
Collelongo, Italy	280.2	1100	11	2
Vielsalm, Belgium	280.2	1000	16	3.4
Camp Borden, Ontario	280.2	814	20	3.7
Tharandt, Germany	280.2	824	16	3.1
Metolius, Oregon	280.8	577	9	3.2
Harvard Forest, Massachusetts	281	1066	18	3.2
Lille Bogeskov, Denmark	281.2	700	13	3.8
Sarrenbourg, France	282.4	820	11	3
Loobos, Netherlands	283	786	20	4.1
Morgan Monroe, Indiana	284.5	1066	8	3.7
Boreaux, France	286.7	900	11	5.6
Walker Branch, Tennessee	287	1355	10	3.8
Duke Forest, North Carolina	287.5	1154	12	5
San Palulo State, Brazil	293.5	1313	10	3.4
Slash Pine, Florida	294.9	1330	8	7
Reserva Jara, Brazil	300.1	2152	8	7.8
Cuieriras, Brazil	300.9	2200	8	8.2

MAP = mean annual precipitation, Rh = heterotrophic respiration, OC = soil organic carbon.
Source: from Sanderman et al. (2003).

Second, water moves pervasively downward in a liquid form, moving dissolved organic molecules and clay-organic complexes downward. This process is akin to advection, a process in which material is transported by bulk motion. Thus, these profiles can be viewed as being the result of a slow pervasive downward movement of SOC accompanied by ongoing decomposition of that C. As will be shown later, there is strong evidence from radiocarbon analyses to support this conceptual framework.

An expression that encompasses all of these processes is[23]

$$\frac{dC}{dt} = -D\frac{d^2C}{dz^2} - v\frac{dC}{dz} - kC + f \tag{7.14}$$

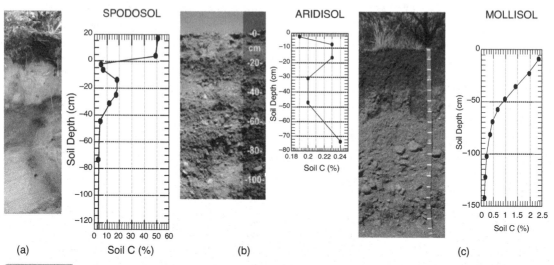

Representative photos of typical soils compared with typical C distributions for these types of soils. Photos of Spodosol (R. Schaetzel), Aridisol (United Emirate Republic), and Mollisol (University of Idaho). Data for soil C from USDA Soil Taxonomy (1975).

where $= -D\frac{d^2C}{dz^2} =$ diffusive movement of SOC, $-v\frac{dC}{dz} =$ advective movement of SOC, $kC =$ decomposition of SOC, and $f =$ plant inputs.

Mathematical software can be used to create flexible time dependent versions of Eq. (7.14). Here, to focus on simple introductions to these concepts and how they can help us better derive process information from soil data, we focus on a simpler steady-state diffusion model:

$$\frac{dC}{dt} = 0 = -D\frac{d^2C}{dz^2} - kC + f \tag{7.15}$$

The analytical solution for this is also somewhat challenging, but a solution has been presented as[24]

$$C(z) = \frac{f}{\sqrt{kD}}e^{z\sqrt{k/D}} \tag{7.16}$$

First, this expression is used to explore how it relates to observed SOM depth profiles. As mentioned earlier, the decomposition rate can be estimated in a number of ways. The key term to parameterize is the diffusion coefficient, D – which was discussed earlier.

Equation (7.16) is for total C. A similar equation can be made for radiocarbon, providing a way to examine how soil age vs. depth should vary. Radiocarbon data, and ages, are reported in a number of ways. One useful method is to compare the ratio (R_{14}) of $^{14}C/^{12}C$ in soil (and compare it with a standard):[25]

$$\text{Radiocarbon age(y)} = -8033 \ln(R_s/R_{std}) \tag{7.17}$$

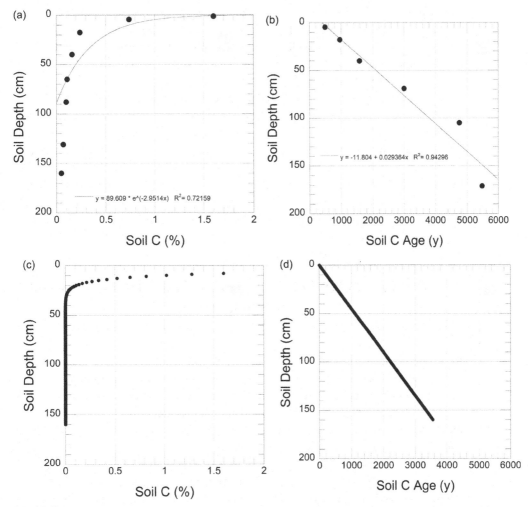

Figure 7.5 (a) Soil C and (b) radiocarbon age vs. depth for a 200 Ky soil in the San Joaquin Valley of California. (c) Modeled soil C content and (d) modeled radiocarbon age vs. depth using Eq. (7.12) parameters derived from Baisden et al. (2002).

R = $^{14}C/^{12}C$ of soil sample (normalized to a stable C isotope ratio) and standard. It is common to correct the sample for the date it was sampled (relative to t = 0, set at 1950). Simply, the steady-state equation for ^{14}C is like that of C plus the radioactive decay constant ($\lambda = 1/8033$ y) and assuming that the differences in diffusivity are negligible:

$$^{14}C(z) = \frac{f}{\sqrt{(k+\lambda)D}} e^{z\sqrt{(k+\lambda)/D}} \tag{7.18}$$

In Figure 7.5, the parameters just discussed are used with Eq. (7.16) to examine soil C content vs. depth. The data are from a 200 Ky-old soil in the grasslands of the Central

Valley of California. The measured soil data show that soil C exponentially declines with depth and that there is a roughly linear increase in the age of SOC with depth (Figure 7.5a, b). Equation (7.16) predicts such a trend in C with depth (Figure 7.5c). Additionally, the radiocarbon model in Eq. (7.18) predicts a linear increase in SOC age with depth, which is what the measured data reveal (Figure 7.5d).

Several caveats should be introduced at this point. First, the assumption that SOC is just one pool likely contributes to the inability to exactly replicate measured values, though challenges exist if multiple pools are also considered.[26] Through estimating as many parameters as possible, and through iterative parameter fitting, multipool models of Eq. (7.16) and (7.18) might better explain observed data and provide insight into the size and cycling of different pools in the soil. These steps are beyond this introductory discussion, but examples of how to do so are found in these papers.[27] Also, it must be recognized that not all soils have these depth trends and that simple box models might be more appropriate for other locations on Earth.[28]

7.6 Soil Carbon Pools and Factors Controlling Soil Carbon Decomposition

Research on the factors and processes that cycle soil C has accelerated greatly in the past two decades as science tries to understand soil's role in the global C cycle and in the evolving human-induced changing climate. It is apparent from a number of approaches that soil C is not one homogeneous pool, but no straightforward means of physically separating these pools without inducing experimental artifacts has emerged. The reason for this is that soil C is physically and chemically heterogeneously distributed with in a soil structural unit (ped), and attempts to dislodge and take samples to the lab for subsequent organic extraction induces uninterpretable changes in the physical arrangement of the C and how we interpret extraction techniques. More simply: C that is physically unavailable to soil microbes under natural field conditions due to its location in the maze of soil particles may become easily accessible after its removal from the profile and after subsequent processing.

This effect is illustrated in a recent study.[29] In that study, samples from a native grassland and an adjacent citrus orchard in California were taken to the lab, and the samples were lightly crushed to remove roots. Then, the samples were incubated so that rates of decomposition and the radiocarbon content of the respiration could be measured (Figure 7.6). The hypotheses were that soil respiration would show the typical initial spike (long considered by microbiologists to be the rapid decay of labile C and freshly deceased microbes) and that the radiocarbon age of the respiration would increase over time as the labile components degraded and as SOM became a more important source for the respiring organisms. The hypothesis about the initial high CO_2 proved correct, but to the surprise of the researchers, this initial flush of soil respiration was up to 13,000 radiocarbon years old. This indicated that, in these soils, the bulk of SOC was simply

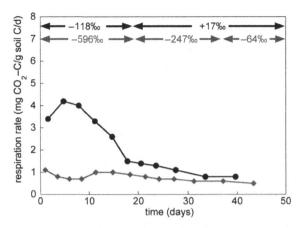

Figure 7.6 Rates of soil respiration vs. time for lab incubations of soil from the A and Bt horizon of a grassland soil in the Sierra Nevada foothills of CA. The bars at the top provide the radiocarbon content of the aggregated CO_2 collected during these time intervals. From Ewing et al., Role of large-scale soil structure in organic carbon turnover: Evidence from California grassland soils, *Journal of Geophysical Research*, 111, doi: 10.1029/2006JG000174 (2006).

physically locked up in macroscopic soil structure, which was fundamentally disrupted and eliminated by simple laboratory processing. Thus, this experiment reveals that most soil C fractionation techniques are being conducted on samples that differ greatly from their nature *in situ*. Quantifying the size and cycling rates of the continuum of soil C pools, while recognizing the physical limitations to its laboratory investigation, remains a major research problem.

Probably no one SOC researcher would be fully satisfied with using the introductory outlines here for a study that informs policy or management. However, these tools, combined with additional steps available from the research literature, form the basis for much of the literature on global soil C. Most importantly, the fact that the processes embedded in these models provide a first-order explanation of the physical controls on soil C storage and its vertical distribution is satisfying for an introductory level of soil biogeochemistry.

7.7 Soil Respiration and CO_2 Transport

Thus far, we have examined the inputs of soil C and rates and patterns of its storage. We have implicitly considered that the CO_2 produced is the loss term (Eq. (7.12)) without considering the processes that actually remove it from the soil. Soil gas exchange, and especially CO_2, is a well-studied area of research, and here the basics of this process are reviewed.

When CO_2 is released by respiring soil microorganisms or by plant roots, it enters the soil atmosphere. The atmosphere is the pore space in soils, a complex and interconnected set of pathways that connect with the overlying atmosphere. The addition of water from rainfall reduces this air space. Thus, the transport of gas in soil – compared with the free atmosphere – must account for pore space, the "tortuosity" or complexity of the pathways, and the water content.

The movement of CO_2, O_2, and other gases in soil is largely by diffusion, the random movement of molecules, but in a net direction dictated by a concentration gradient:

$$\varepsilon \frac{dC}{dt} = D_s \frac{\partial^2 C}{\partial z^2} + \phi \tag{7.19}$$

where ε = free air porosity, C = CO_2 concentration (e.g. mol cm^{-3}), D_s = effective diffusion coefficient in soil (cm^2 s^{-1}), ϕ = CO_2 production (mol cm^{-3} s^{-1}), and z = depth (cm).

Based on the ranges in the value of D_s, soils likely achieve steady state in less than a day after a major perturbation, so it is common to assume steady-state conditions, which allows Eq. (7.19) to be solved:[30]

$$C = \frac{\phi}{D_s}\left(Lz - \frac{z^2}{2}\right) + C_{atm} \tag{7.20}$$

The diffusion coefficient for CO_2 in air is about 0.14 cm^2 s^{-1}. A number of models have been published to adjust this value to soil conditions. To do so, one usually needs information on bulk density, gravel content, and water content. Here, as an example, is a model that incorporates these approaches:[31]

$$D_s = D_{atm} \frac{\varepsilon^{2.5}}{\sqrt{\gamma}} \tag{7.21}$$

where D_o = diffusivity of gas in air, D_s = diffusivity in soil, ε = air filled porosity (cm^3 cm^{-3}), and γ = total soil porosity (cm^3 cm^{-3}). Values for γ and ε are calculated as

$$\gamma = 1 - \left(\frac{\rho_b}{\rho_p}\right) = \varepsilon + \theta \tag{7.22}$$

where ρ_b = bulk density, ρ_p = particle density (assumed at 2.65 g cm^{-3}), and θ = volumetric water content.

While useful and widely used, Eq. (7.20) is a simple beginning point for more complicated models that may include varying soil CO_2 production rates with depth, varying values of the diffusion coefficient with depth, and, of course, removing the assumption of steady state. The illustrations that follow show that Eq. (7.19) does provide a realistic first-order understanding of soil gas exchange. At soil depth $z = 0$, the flux of CO_2 from the soil to the atmosphere (J_{CO_2}) is

$$J_{CO_2} = -D_s \frac{\partial C}{\partial z} \tag{7.23}$$

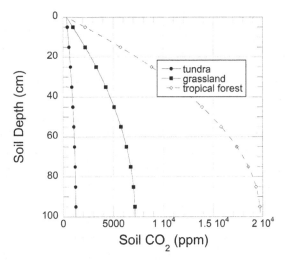

Figure 7.7 Calculated soil CO_2 profiles (Eq. (7.20)) for three ecosystems with measured rates of respiration. Parameters given in text.

The flux of CO_2 across the soil surface/atmosphere boundary is called soil respiration and is commonly measured with chambers installed on the soil surface. Compilations of soil respiration rates from around the world have been made, and the environmental literature contains many more individual studies.[32] Here, respiration from three distinct environments is considered: tundra (60 g C m^{-2}y^{-1}), grassland (442 g C m^{-2}y^{-1}), and tropical forest (1260 g C m^{-2}y^{-1}). By knowing the respiration values, one can use Eq. (7.15) to calculate the soil CO_2 depth profiles. Here, assuming a D$_s$ value of 0.0342 cm^2 s^{-1} and an atmosphere of 350 ppm (Figure 7.7), as respiration rates (and rates of soil CO_2 production) increase, the soil CO_2 concentrations increase. Conversely, from measured concentration profiles, one can use Eq. (7.20) to calculate the flux rate to the atmosphere, which increases with increases in the concentration gradients in the soil (Eq. (7.23)).

The transport of CO_2 from soil to the atmosphere essentially completes the soil C cycle, and it also has important impacts on other areas of soil biogeochemistry. First, CO_2 dissolved in water forms carbonic acid, and increased partial pressures of CO_2 increase the acidity available for mineral weathering, which thus should generally increase with CO_2 concentration. In many soils, as a result of carbonic acid weathering of silicate minerals, the dominant anion is HCO$_3^-$, which is derived from CO_2 and which represents a sink for atmospheric CO_2. One important solid sink for CO_2 in arid and semiarid soils is the mineral calcite (or also carbonate):

$$(Ca^{+2}) + 2(HCO_3^-) = CaCO_3 + 2(H^+) + CO_2 \qquad (7.24)$$

The chemistry of carbonate equilibria in water is a complex topic unto itself,[33] but simply when the product of $(Ca^{+2})(CO_3^{-2}) > K_{eq}$, carbonate minerals are favored to precipitate in soil. Because calcite is a moderately soluble mineral, K_{eq} is not commonly exceeded in humid environments where soil water concentrations of ions remain relatively low. In contrast, drier environments have periods of time when water is removed by evapotranspiration, thus causing the equilibrium constant of calcite to be exceeded. The depth at which calcite begins to form in soils is, all else being equal, a function of rainfall: as rainfall increases, the depth to calcite increases, and it ultimately disappears at mean annual rainfalls above 50 to 100 cm y^{-1}.[34]

One of the unique and informative properties of soil calcite is that it chemically retains information about the rates of soil CO_2 production, and the types of vegetation, under which it formed. This has proved to be a useful tool in using ancient and buried soils to reconstruct a number of aspects of Earth's atmospheric history, vegetation history, and climate history. Here, we will not review those studies but instead extend our introduction of CO_2 transport to explicitly include the two stable isotopes of C: ^{12}C and ^{13}C. Isotopes are species of elements that have the same number of protons and electrons but differing numbers of neutrons (hence the difference in atomic mass number). The isotopes of many elements have roughly similar chemical behavior, but due to differences in mass, they react and are transported at different rates. ^{12}C is by far the most abundant isotope of C on earth (98.9 percent), while ^{13}C is relatively rare (1.1 percent). The radioactive isotope ^{14}C is exceedingly rare (about 1‰).

Cerling[35] introduced the concept that in soils, the ratio of ^{13}C to ^{12}C in soil CO_2 (and in the calcite that forms from it) can be described by the ratio of two production diffusion equations: Eq. (7.20) as the denominator, assuming that bulk C approximates ^{12}C, and the numerator an equation for ^{13}C:[36]

$$R_s^{13} = \frac{\left(\frac{\phi R_p^{13}}{D_s^{13}}\right)\left(Lz - \frac{z^2}{2}\right) + C_{atm}R_{atm}^{13}}{\left(\frac{\phi}{D_s^{12}}\right)\left(Lz - \frac{z^2}{2}\right) + C_{atm}} \tag{7.25}$$

where R_s, R_p, and R_{atm} refer to the isotopic ratios of soil CO_2, plant carbon, and the atmosphere. D_s13 = the diffusion coefficient in soil adjusted for ^{13}C: $D_s/1.0044$.

While mass spectrometers measure the isotope ratios of compounds, they are commonly reported as δ^{13}C values:

$$\delta^{13}C(‰) = ((R_s/R_{std}) - 1)1000 \tag{7.26}$$

The value of the ^{13}C/^{12}C ratio in the international standard (std) (a marine fossil) is 0.0112372.[37] Common δ^{13}C values for parameters in Eq. (7.17) are: plants = −27‰ for C3 plants, −14‰ for C4 plants, and about −7‰ for the atmosphere[38](which is constantly becoming more negative through the burning of −27‰ fossil fuel). An inspection of Eq. (7.25) shows that for a given set of parameters, the ratio of ^{13}C/^{12}C is dependent on the rate of CO_2 production. For the three soils examined in Figure 7.7, their approximate δ^{13}C values vs. soil depth are examined assuming that they all have the same plant composition (Figure 7.8). The results show that as CO_2 rates increase, the C isotope values approach those of the plant source (plus a 4.4‰ enrichment due to the fact that ^{13}C diffuses more slowly than ^{12}C). This is due to the fact that as production rates decline, the impact of the atmosphere increases to greater depths in the soil (see Eq. (7.25)). It should be noted that, while the steady-state soil CO_2 becomes enriched in ^{13}C, the steady flux out of all these soils must equal that of the plant source (−27‰) in order to retain mass balance.

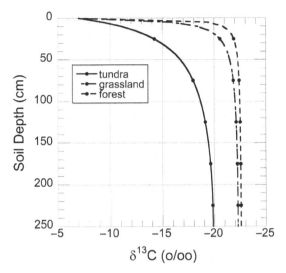

Figure 7.8 Calculated soil CO_2 profiles (Eq. (7.25)) for three ecosystems with measured rates of respiration. Parameters given in text.

One example illustrates the importance of these relationships for using profile soil data to learn about past climate conditions. Ebeling et al.[39] measured the $\delta^{13}C$ values of soil carbonate in an ancient soil in northern Chile, where the enormous age of the soil (at least 2+ million years) means that it had begun soil formation and carbonate accumulation before a profound aridification started at the end of the Pliocene period: climate change due to changes in Pacific Ocean circulation. At equilibrium, calcite is about 10‰ enriched in ^{13}C relative to the CO_2 it forms from.[40] The present soil CO_2 production rates are almost 0 due to the extreme aridity (MAP = ~20 to 25 mm y^{-1}). The measured $\delta^{13}C$ values of the carbonate minerals (assuming that the flora has always been C3 vegetation) require a soil profile CO_2 isotope composition resulting from soil respiration rates that presently occur in environments with rainfall rates of 100 to 300 mm y^{-1}. This provides some guidance to the magnitude of aridification the region has experienced over time and shows how soils can retain information from the distant past.

It should be noted that Eqs (7.20) and (7.25) (with appropriate modifications) are applicable to the transport (by adding consumption terms) and isotope composition of N_2O or other trace gases with additional source or sink terms. The interested reader can use the basic principles developed here to further explore these questions.

7.8 Summary

Plants and animals are mainly composed of reduced organic compounds that upon oxidation, mediated by soil microorganisms, produce CO_2, heat, and organic matter.

We are all largely "fire and air," as Cleopatra proclaimed, mourning the death of Antony. One of the earliest discoveries by pedologists is that the soil C pool, to which we are eventually linked, is predictably distributed across the globe in a pattern closely related to climate – a pattern that is describable using simple tabulations of inputs and outputs from the soil.

This pool of organic C plays an increasingly important role in our present world. Simply dividing the total amount of C in atmospheric CO_2 by the annual rate of soil respiration shows that an average CO_2 molecule in the atmosphere passes through soil organic matter, somewhere in the world, every 13 years or so.[41] Humanity has now reached a point in history where we can alter this rate (usually speeding it up), although many interesting questions remain about how we have changed these background rates and how we can use our knowledge to better manage this important global C pool. This will be examined in greater detail in the final chapter of this book.

7.9 Activities

7.9.1 Soil Carbon Calculations

A soil has 15 kg C m^{-2} to a depth of 1 m. It has a turnover time (residence time) of 1000 years.

a. What is the decomposition constant (and units) (assuming near steady-state conditions)?
b. What is the C input rate (and units)?

7.9.2 Soil Carbon Dynamics

For the data in Section 7.9.1, using the appropriate equation from the chapter (and assuming no C at t = 0), plot the C content of the soil for 5000 years. When does it become roughly time invariant?

7.9.3 Soil Carbon Steady State

For the soil examined in Sections 7.9.1 and 7.9.2, calculate and plot the steady-state C content vs. soil depth. Assume that the diffusion coefficient for C movement is 0.5 cm^2 y^{-1} and that the soil has a constant bulk density of 1.4 g cm^{-3} (the key is to make sure units are correct). Present the results as percent C vs. depth.

7.9.4 Soil Greenhouse Gas Calculation Primer

When a soil is at steady state (meaning the soil is time invariant), in any soil layer, the concentration of CO_2 is the balance between diffusion inputs into and out of the layer, plus the production (or minus the consumption). Soil processes do indeed change with time, but the steady-state assumption is a simple starting point for calculations.

The movement of gas is by diffusion. Diffusion is movement along a concentration gradient, dC/dz, times a constant for that gas. Molecular diffusion is relatively slow and is made even slower by reduced free air volumes in soil. The simplest starting equation for diffusion is

$$J_s = -D_s(dC/dz)$$

Where J_s = flux of CO_2 across the soil/air interface (or some boundary) (moles CO_2 /cm^3 s^{-1}), D_s = cm^2 s^{-1}, C = moles cm^3, z = cm. The minus sign indicates that the soil is losing CO_2 if it is being transferred toward the atmosphere.

Soil greenhouse gas concentrations are commonly given in ppm (micromoles of gas per mole of dry air). The units in flux calculations require moles per volume. Thus, we use the ideal gas law (PV = nRT) to make the conversion. For example, for air at 410 ppm CO_2, the number of moles per cm^3 is

$$n = -PV/RT$$

where V = 1 cm^3; R = 82.05746 cm^3 atm K^{-1} mol^{-1} (the gas constant, for the combination of units used here, is available online from a number of sites, including Wikipedia); T = 273 + 25 C = 298 K; P ~ 410 ppm = 410/1 000 000 = 0.00041 atm.

Thus,

$$n = 1.677 \times 10^{-8} \text{ mol } CO_2 \text{ (per cm}^3 \text{ air)}$$

The diffusivity of CO_2 in air is 0.16 cm^2 s^{-1}. This is reduced in soil due to reduction in air volume by solids and water. If the soil has a bulk density (BD) of 1.4 g cm^3, then the total porosity is

$$\text{Porosity} = 1 - (BD/PD) = 1 - (1.4/2.6) = 0.46$$

Where if the soil has 10 percent by volume water (0.1), then the air filled porosity = 0.46 − 0.1 = 0.36. Using the following expression, we can thus calculate the D_s from D in pure air:

$$Ds = 0.16 \times (0.362.5/0.460.5) = 0.0084$$

For the soil profile below, if the first layer has 1000 ppm CO_2, the atmosphere has 410 ppm, and the gradient is 10 cm, the flux rate is

$$J = -D(dC/dz) = -0.0084[(4.08946 \times 10^{-8} - 1.67668 \times 10^{-8})/10] =$$
$$-2.035978 \times 10^{-11} \text{ mol } CO_2/\text{cm}^2 \text{ s}^1 = -6.42 \text{ moles } CO_2/\text{m}^2 \text{ y} =$$
$$-77.0 \text{ gC/m}^2 \text{ y}$$

This calculation can be performed for each soil layer (e.g. IN-OUT) to determine how the production rate of CO_2 changes with soil depth, since the difference between what diffuses into a layer and what diffuses out must be the production or consumption.

There are various reasons why one might want to explore the range of potential CO_2 concentrations with depth. For example, ecologists have measured CO_2 emissions from soils in many locations, but how the concentration varies in the soil is not measured. This is where the steady-state depth model can be used:

$$C = \frac{\phi}{D_s}\left(Lz - \frac{z^2}{2}\right) + C_{atm}$$

where L = total depth of soil being considered, z = specific depth, D_s= soil diffusion coefficient, ϕ = symbol for CO_2 production rate (mol CO_2 cm^{-3} s^{-1}), C_{atm} = atmospheric concentration (mol cm^{-3}).

7.9.5 The Key Gases

Microorganisms in the soil produce and/or consume the three important greenhouse gases in the atmosphere: CO_2, CH_4, and N_2O.

CO_2 is produced as the product of respiration, the oxidation of reduced C compounds with atmospheric O_2. It diffuses to the atmosphere by the process of diffusion. CO_2 production is proportional to organic C, and we are also measuring the organic C and N content of this soil. The current atmospheric concentration is about 409 ppm.

CH_4 is largely in the atmosphere due to production in low-oxygen conditions (lake sediments, leakage from oil and gas wells, cow stomachs, etc.). In aerobic soils, it is consumed by organisms called *methanotrophs*, bacteria and archaea that metabolize the CH_4 to other C compounds as their only source of energy. Some methanotrophs can oxidize methane in anaerobic conditions using NO_3 as an electron acceptor. The transfer of atmospheric CH_4 into soils is also caused by diffusion. Atmospheric concentration is approximately 1.859 ppm.

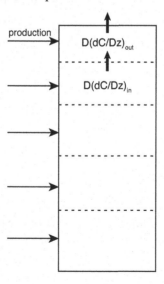

N_2O (nitrous oxide) is one of three N gases that can be produced in the complex cycling of N by soil microorganisms. It can be produced during both aerobic and anaerobic conditions. The concentration of N_2O in the atmosphere has increased due to fertilization with N, which is then converted by microbes into N_2O. Yet, it is hard to predict the concentration and flux rates of N_2O. One of the reasons may be the recent discovery that there is a large group of microbes that contain genes for producing the enzyme that reduces N_2O to N_2 (a non-greenhouse gas that makes up 80 percent of the atmosphere). Thus, there is likely both production and consumption of N_2O occurring simultaneously. Therefore, we will see whether our soil is a net producer or a small consumer of the N_2O in the atmosphere. The atmospheric concentration is about 0.330 ppm or 330 ppb.

7.9.6 Flux Calculations

For an oak-grassland soil on the Berkeley campus, in the fall after the long summer dry season, the following data were obtained:

Depth (cm)	N_2O (ppm)	CH_4 (ppm)	CO_2 (ppm)
0	0.31518034	1.8186948	474.77227
20	0.31081714	1.6425408	1363.54437

(Note that the 0 depth values do not exactly match that of the well-mixed atmosphere.) Calculate the flux rates of the gas exchange between the soil and the atmosphere using the concentrations and the following information to calculate diffusion coefficients:

Diffusivity of gases in air	D (cm^2 s^{-1})
CO_2	0.138
CH_4	0.196
N_2O	0.143

Soil bulk density = 1.5 g cm^{-3}
Rock bulk density = 2.6 g cm^{-3}
Water content ambient = 0.1 cm^3 water cm^{-3} soil
Soil porosity = 1 − (bulk density soil/bulk density rock)
Air fill porosity = total porosity − water volume
$D_s = D_{atm}$(air filled porosity$^{2.5}$/total soil porosity$^{0.5}$)

For the soil examined in Sections 7.9.1–7.9.3, use the diffusion coefficient calculated in Section 7.9.4, assume the loss of C as CO_2 is the same as C inputs, and assume that the production of CO_2 is constant with depth (e.g,. total production divided by total distance, which we will assume is 100 cm in this example). Calculate the CO_2 concentration vs. soil depth and report the results in ppm.

The C isotope composition of the soil CO_2, in regions where $CaCO_3$ forms, is reflected in the C isotope composition of the soil minerals. In many arid regions, this profile is an integrated average of the biological productivity at a site. In the chapter, an equation (Eq. (7.25)) for calculating the C isotope composition vs. depth was introduced, but it involves converting C isotopes in the delta notation to ratios and then back again. A simpler (and correct) alternative is to calculate the C isotope composition vs. depth from a simple mixing model:

$$\delta^{13}C_{CO2} = [(\delta^{13}C_{atm})([CO_2 \text{ atm}]) + (\delta^{13}C_{plants})([CO_2 \text{ soil}} - CO_2 \text{ atm}])]/[CO_2 \text{ soil}]$$

$$\delta^{13}C_{atm} = -8\text{‰}$$

$$\delta^{13}C_{plants} = -27\text{‰} + 4.4\text{‰ (diffusion enrichment)} = -22.6\text{‰}$$

Plot the C isotope values of the soil CO_2 profile vs. depth. How does the nature of this profile change when respiration rates decline, and what would signify a soil that has low rates of biological productivity?

8 Chemical and Physical Processes in Soils

The processes of heat, gas, and water exchange (and the chemical reactions that they drive) are the processes that make soil "soil." These transfer rates are controlled by the boundary conditions of the soil system, which are defined by the system's state factor configuration. While we have shown that correlating soil properties with variations in state factors is both informative and very useful, it is important to probe the processes that occur in the soil system and gain some insights into understanding them.

In this chapter, transfers are examined in one dimension (vertical). Later, when we examine hillslopes (Chapter 9), horizontal transfers of soil along slope gradients will be evaluated. The vertical fluxes of mass and energy are what largely make the various types of soil horizons (combined with biophysical mixing processes). The formation of some, though not all, of these horizons has been adequately described by transport models. There are sophisticated numerical reaction-transport models in the hydrogeosciences that are being adapted and applied to describing various soil processes. Here, in what is intended to be an introduction to better visualizing and conceptualizing vertical processes, a simpler approach is taken. Some relatively simple numerical and analytic models are discussed that help one begin to think more deeply about the processes controlling soil biogeochemistry.

8.1 Transfer of Heat, Gas, and Water in Soil

Heat, gas, and water in soil move along gradients that represent how those entities change with distance. For gas movement, discussed in Chapter 7, it is commonly how concentration (mol cm^{-3} of soil air volume) changes with vertical distance. For water movement, it is the change in water content or water potential vs. distance. For heat flux, it is a temperature gradient. Mathematically, all these transport processes have the form of:

$$q = -K \frac{\partial quantity}{\partial depth} \tag{8.1}$$

where q = flux or flow rate (some quantity vs. time), and K is a constant that reflects the ability of the soil to transmit the entity of interest.

Here, the relationships of this equation to quantifying chemical weathering and weathering rates are introduced.

8.2 Chemical Weathering of Minerals

Chemical weathering is the alteration of minerals by aqueous solutions. The mechanisms by which this is accomplished are varied: H^+ can displace cations in a silicate mineral structure, OH^- can lead to high pH values and changes in the solubility of silicate minerals, and oxidation-reduction reactions can impact elements with multiple oxidation states. As discussed in Chapter 2, mineral weathering rates are reported based on a measure of surface area, which emphasizes the importance of water–mineral interactions and how the mechanisms described earlier are largely restricted to surface processes. In addition to surface area, other parameters in reaction models account for temperature (which is critical to any biological or abiotic reaction) and the nature of the waters (such as pH). The surface reactions, the rate laws, and the thermodynamics of these processes are covered in many textbooks on soils and natural waters.[1] Here, we focus on the observable field-scale effects of weathering and chemical reactions, particularly on the gain, loss, and mobility of elements in differing spectrums of the climate continuum.

Chemical weathering releases soluble components to the soil water. In the field, studying the rates of chemical weathering can be approached either by monitoring the flux and concentration of the soil water (which reflects the solutes released by the minerals) or by examining the solid phase itself to detect the gains and losses of elements and changes in mineralogical makeup. This chapter focuses on the latter. Solid-phase analyses of soils is readily available and offers both students and researchers a number of opportunities for quantitative research analyses. It should be noted that the soil water–based approach is indeed used in a number of studies and even as a complementary approach to solid-phase studies.[2]

In humid environments where water flows through soil, reacts with minerals, and removes weathering products, soils eventually collapse due to the loss of mass from these processes (Figure 8.1). Some elements are concentrated (if they are insoluble), and some are almost entirely removed and depleted. Some are transferred to lower horizons. One can quantify both the mass gain/loss and the volumetric change in any and all soil horizons using total chemical analyses, bulk density measurements, and a mass balance model that has evolved over the years.[3] In this approach, the gain or loss of an element (or total mass) in a soil horizon relative to the starting parent material can be calculated by normalizing the concentration of the mobile element (j) of interest (for example, Ca) to an element that is essentially chemically immobile (i) in the soil weathering environment (Zr, Ti, etc.). This ratio ($R_s = \frac{c_{j,s}}{c_{i,s}}$) is then compared with that of the parent material ($R_p = \frac{c_{j,p}}{c_{i,p}}$) to calculate the fraction gain (+) or loss (−) of that element relative to the parent material:

$$\tau = \frac{R_s}{R_p} - 1 \tag{8.2}$$

Loss (or tau) values maximize at −1 (e.g. −100%), while gains can be much higher than 1 (or 100 percent).

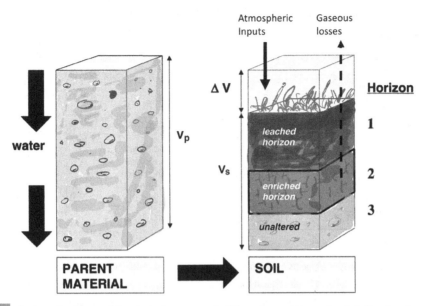

Figure 8.1 A schematic of some of the physical processes of soil formation and biogeochemical alteration. A parent material of a given volume (V_p) is reacted upon by water and acids, losing mass and volume (ΔV), leaving behind largely leached (or depleted layers) and some layers that may have experienced partial gains of some elements (V_s).

What constitutes an immobile element? To better understand this, it is recommended to use the paper entitled "An earth scientist's periodic table of the elements and their ions" as an important reference.[4] In that paper and its illustrations, the elements are color coded and arranged according to their solubility in soil weathering conditions. One key quantity that impacts the solubility of an element in water is its *ionic potential,* which is the ratio of the electrical charge of an ion to its radius (in angstroms) (z/r). In general, ionic potentials of 2 or less (most of the alkali metals and alkaline earths) form weak bonds with O and tend to be relatively soluble in water. Elements with z/r ratios of 2 to 8 strongly attract O and tend to form insoluble oxides in soils (Figure 8.2a). Finally, elements with z/r ratios >8 also strongly attract O but tend to form soluble oxyanions in solution (such as SO_4^{-2}) (Figure 8.2b).

Commonly, elements in groups 4 and 5 of the periodic table are candidates for immobile elements based on these principles, and out of these, Zr and Ti have likely been used the most frequently in mass balance calculations.

However, in addition to being immobile in the weathering environment, candidate elements must be homogeneously distributed throughout the parent material. In igneous rocks, for example, Zr or Ti would likely be evenly distributed. However, in sedimentary rocks or, more specifically, alluvial deposits, heterogeneous distributions of elements are common due to particle sorting during sedimentation. Rutile (a source of Ti) and zircon (a source of Zr) are heavy minerals: rutile = 4.25 g cm^{-3} and zircon = 4.65 g cm^{-3}. During fluvial transport these minerals are differentially sorted and concentrated by differences in water velocity and thus can be differentially concentrated in lenses or by depth in fluvial deposits. Choosing an appropriate index element in these types of soils is more challenging. Gromet and Silver[5] demonstrated that

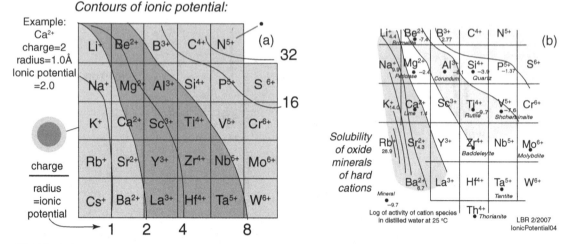

Figure 8.2 (a) Arrangement of elements on the periodic table by ionic potential, with the relatively water-soluble elements with ionic potentials <2 and >8, and the immobile elements in the intermediate range of 2–8. (b) As in (a), but illustrating the log of activity (essentially concentration) in water of some minerals containing specific elements.

another group of relatively refractory elements, the lanthanides or the rare earth elements, tend to be distributed through all the major silicate minerals in low abundances and thus may be better immobile index elements than Ti and Zr in soils derived from fluvial or other transported material.[6] For example, in Chile, in a Miocene-aged soil formed on an alluvial fan, the soil had an erratic and unpredictable correlation of Ti and Zr (both found in heavy minerals) in both the <2 mm and the gravel fractions (Figure 8.3a). In contrast, correlations between Nd and Ce (both lanthanides) were linear in all the soil horizons, and thus these were used as the index elements in that particular study.

There are two other key calculations that are important to quantify soil changes over time or space: the total mass loss or gain (the preceding equation gives the fractional, not mass, loss) and the volumetric changes of the soil. The absolute gains or losses of an element in mass per unit volume of the parent material ($\delta_{j,s}$) can be calculated using

$$\delta_{j,s} = \frac{\tau C_{j,p} \rho_p}{100} \tag{8.3}$$

where ρ_p is the bulk density of the parent material. The volumetric change in the soil, or strain (ε), can be calculated as

$$\varepsilon_{i,s} = \frac{\rho_p C_{i,p}}{\rho_s C_{i,s}} - 1 \tag{8.4}$$

where ρ_s is the bulk density of the soil horizon.

Total mass loss or gain for the soil profile, as well as total volumetric change, can be determined by summing the results of Eqs (8.3) and (8.4) for the entire soil profile.

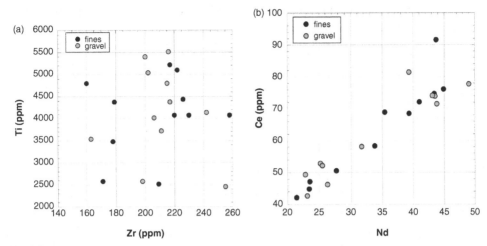

Figure 8.3 An example of how immobile element candidates may behave in fluvially sorted sediment. (a) erratic correlation between Ti and Zr (in both the gravels and the <2 mm fraction of the soil horizons) and (b) the tight correlation between two lanthanides, Ce and Nd. Due to their presence in minerals of all types, both Ce and Nd were considered good immobile elements for the soil being considered.

| Box 8.1 | Sources of Soil Chemical Data |

In the past few decades, the use of total soil chemical analyses has regained interest among soil scientists and geochemists interested in understanding rates of chemical weathering and biogeochemical cycling in soils *in situ*. To obtain a full suite of elements (major elements, trace elements, and rare earths) usually involves several methods of analyses: chemical dissolution followed by inductively coupled plasma atomic emission spectroscopy (ICP-AES) or ICP-AES analyses for the major and trace elements. Besides university laboratories, private labs are readily available, which perform these analyses for a number of geochemical and pedological clients.

Data are available in the scientific literature (which is accessed in this book) and also on the Web. The National Resources Conservation Service (NRCS) has begun to conduct total chemical analyses more frequently alongside its traditional soil characterization methods. The locations and data can be accessed from the NRCS Soil Characterization website (https://ncsslabdatamart.sc.egov.usda.gov). In addition to the USA, soil data is also available in a more limited quantity from a number of other nations. Additionally, the United States Geological Survey (USGS) has sampled and analyzed over 4800 soils in the USA, and the data and interactive maps are available on the Web (https://mrdata.usgs.gov/soilgeochemistry/#/summary).

To illustrate the use and importance of these principles, data from a study of the marine terraces at Santa Cruz, California, by White and many coauthors is examined.[7] The terraces of beach sand, gravel, and eolian sand overlie wave-cut benches on mudstone and sandstone. Chemical data for Na, Ca, and Ti for the lowest horizons of the youngest soil, and for the soil on terrace 2 (90 Ky based on cosmogenic radionuclide dating methods[8]), are presented in

Table 8.1. From this, an average parent material value for mobile and immobile elements was obtained and used to calculate the τ, ε, and δ values for the soil (Figure 8.4a–f). Here, the steps in the calculations and an analysis of the results are presented.

Table 8.1 Soil chemical data for terrace 2 of the Santa Cruz marine terraces

Location	Depth (m)	Na_2O (%)	CaO (%)	TiO_2 (%)	Zr (ppm)	SiO_2 (%)	Quartz
Lower three horizons, terrace 1	–	2.73	2.05	0.36	76	75.8	42.6
		2.96	2.21	0.32	74	75.6	42
		2.95	2.14	0.31	69	75.2	42.1
Terrace 2	0.25	1.07	0.67	0.49	278	79.6	64.3
	0.4	1.11	0.64	0.49	281	80.6	62.2
	0.56	1.04	0.53	0.5	269	80	67.1
	0.7	0.98	0.49	0.51	249	79.6	58.9
	0.86	0.74	0.35	0.55	211	76.6	55.2
	1	0.86	0.37	0.57	226	77.9	50.1
	1.3	0.79	0.32	0.64	165	74.4	40.5
	1.58	0.96	0.39	0.56	149	74.9	39.3
	1.7	1.28	0.52	0.48	153	76.9	
	1.87	1.37	0.53	0.43	124	76.7	43.9
	1.95	1.49	0.63	0.37	95	76.8	42.6
	2.2	1.66	0.68	0.35	74	76.1	41.8
	2.58	1.79	0.73	0.44	70	73.1	
	2.91	1.99	0.82	0.41	69	73.8	35
	3.37	1.97	0.83	0.43	68	73.6	33.7
	3.6	1.95	0.83	0.43	74	73.5	33.7
	4.06	1.75	0.68	0.86	137	69	26.8
	4.4	2.3	1.48	0.18	60	77.6	42.7
	4.9	2.6	1.83	0.21	75	76.8	42
	5.26	2.67	1.81	0.15	61	77.2	44.3
	5.84	2.65	1.88	0.2	71	76.7	44.9
	6.14	2.69	1.88	0.19	65	76.9	43.2
	6.78	2.59	1.8	0.17	105	77.7	45.3
	7	2.65	1.8	0.2	75	77.5	44.6
	7.22	2.67	1.79	0.17	60	77.3	41.6
	7.62	2.59	1.82	0.16	55	78.2	
	8.84	2.64	1.92	0.16	58	78	
	10.06	2.68	1.91	0.16	65	77.8	
	11.28	2.6	1.83	0.13	59	78.9	
	12.5	2.62	1.82	0.15	62	78.1	
	13.26	2.68	1.89	0.2	72	77.2	
	14.33	2.68	1.9	0.2	73	76.9	
	15.34	2.72	1.92	0.24	71	76	

Data from A. F. White et al., Chemical weathering of a marine terrace chronosequence, Santa Cruz, California I: Interpreting rates and controls based on soil concentration-depth profiles, *Geochimica et Cosmochimica Acta*, 72: 36–68 (2008).

The first step is the choice of parent material and the associated choice of an index element. White et al.[9] used quartz (a highly resistant mineral) as the index element. However, not many studies commonly have the quantitative mineralogical analyses presented in this study, and a comparison of Zr with quartz shows a good correlation; thus, Zr is used as the index element here (Figure 8.4a). Next, the concentration and tau values for Na_2O are presented (Figure 8.4b,c). In this case, both concentrations and tau values show the same patterns with depth: low concentrations or large losses near the surface, a linear increase in concentration and tau values with depth, and then a constant concentration with depth, which reflects the composition of the unweathered sediment. This pattern has significant importance because it reflects the slow movement of soil water through the soil and its chemical reactions that equilibrate with water. It will be shown later that as the water reaches chemical equilibrium with the minerals supplying Na (in this case, the mineral albite), weathering ceases. Thus, the two figures show the weathering front of this slow process. Second, the reason why both concentrations and tau values show the same trend is that Na is present in largely one slightly soluble mineral, and concentrations reflect the loss of that mineral (albite). If one examines another element (SiO_2), it is found that a large difference between concentrations and tau values exists, illustrating why normalization to an immobile index element is required to quantify gains and losses over time. Figure 8.4d shows SiO_2 concentrations with depth. A simple interpretation of this pattern would be that if we assume the lower depths represent unaltered sediment, then the upper 1.5 m or so has gained SiO_2, while depths between 2 and 4 m have experienced some losses. However, this interpretation is flawed because Si is found in both soluble minerals (such as albite) and insoluble minerals (quartz). The slow removal of albite and other minerals, and the resulting concentration of quartz, results in an apparent (but not true) increase in SiO_2. Figure 8.4e shows the tau values for SiO_2: The upper 2 m has lost 60 to 75 percent of its initial SiO_2, and the depth pattern of losses largely reflects that of Na (Figure 8.4c), as would be expected if Na silicate minerals are undergoing dissolution.

Sodium and Ca have different fractional loss rates (Figure 8.4f) with higher fractional losses of Ca vs. Na. White et al. noted that plagioclase minerals tend to preferentially release Ca vs. Na during chemical weathering as a mechanism for this observation. As a result of mass loss, the upper 2 m of the soil has lost more than 50 percent of its original volume, and the landscape has undergone collapse over time (Figure 8.4g, Eq. (8.4)). As Eq. (8.4) indicates, volume change can occur through changes in bulk density relative to the parent material and/or through the loss or gain of mass. Since the bulk densities of the soil horizons are lower than that of the parent material (representing physical expansion of material), the net collapse of the upper part of the soil profile is due to loss of mass due to chemical weathering and leaching of the dissolved solutes. Using Eq. (8.3), one can calculate the mass of Na_2O loss per cm^3 of volume (Figure 8.4h). If these values are divided by soil age, the resulting rates (($g\ cm^{-3}$)/Ky) reveal the average long-term chemical loss rates. If Figure 8.4h is integrated by depth (and converted to moles), a loss of 2085 moles of Na_2O per m^2 of land surface, over the apparent age of 90 Ky, is the resulting rate of loss.

One of the features of Figure 8.4c,e is that the depth profiles of the elements show a distinctive trend: (1) (except for small concentrations at the surface caused by atmospheric

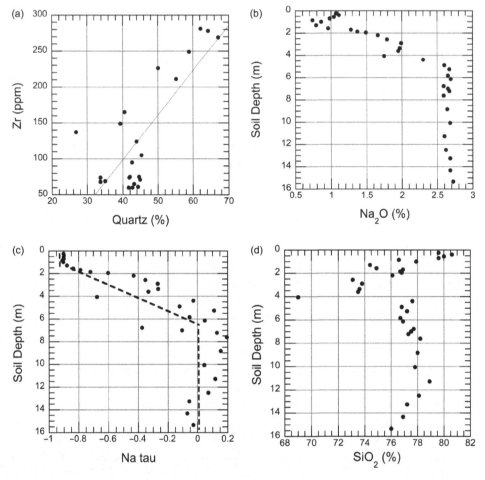

Figure 8.4 Data and calculations for a 90 Ky soil on a marine terrace near Santa Cruz, CA. (a) Zr concentration vs. quartz percentage in all soil horizons, (b) Na_2O concentration vs. depth, (c) the Na tau value (Zr) vs. depth with key slope changes identified with dotted line, (d) SiO_2 concentration vs. depth, (e) the Si tau value (Zr) vs. depth, (f) Na tau vs. Ca tau values (Zr), (g) volumetric change (epsilon value) vs. depth, and (h) gain or loss (g cm^{-3}) of Na_2O vs. depth. All data from Table 8.1.

deposition and biological cycling) a near complete removal from the surface to some depth (1–2 m), (2) a linear increase in concentration over some distance (roughly 2 to 4 m) to (3) parent material values, which are then constant with depth. This depth profile reflects the downward-moving processes of chemical weathering. This is schematically illustrated in Figure 8.5. White et al.[10] presented an expression that describes this pattern:

$$R = \frac{dC}{dz}\frac{1}{S_v}\omega \tag{8.5}$$

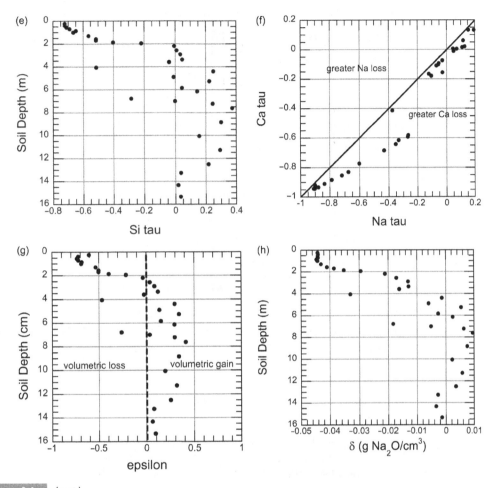

Figure 8.4 (cont.)

where R = mineral weathering rate (mol m^{-2} surface area s^{-1}), $\frac{dC}{dz}$= the concentration gradient (mol m^{-3} m^{-1}), S_v = surface area of mineral per volume of soil in the weathering zone (m^2 m^{-3}), and ω = the weathering front velocity (m s^{-1}). Equation (8.5) can be rearranged to show its similarity to Eq. (8.1):

$$R = K\frac{dC}{dz} \tag{8.6}$$

where K = $\frac{1}{S_v}\omega$ and has units of s^{-1}, a velocity coefficient. The units are chosen to correspond to typical units used to denote mineral weathering rates. Additionally, the term $\frac{dC}{dz}$ is used here (the inverse of the term b_s used by the original authors) to emphasize that the rate term is related to a gradient in concentration (as well as other terms).

The weathering front velocity (ω) is driven by the amount of water moving through the soil (q_s) (which drives chemical weathering) and the mass of mineral dissolved per volume of soil water ($m_{solution}$) relative to the initial mass in the soil parent material (M_{total}):

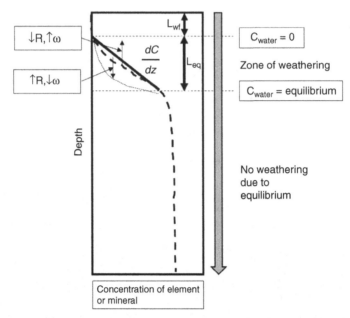

Figure 8.5   A schematic representation of ways to interpret a soil weathering profile. Based, in part, on K. Maher, The dependence of chemical weathering rates on fluid residence time, *Earth and Planetary Science Letters*, 294: 101–110 (2010).

$$\omega = q_s \frac{m_{solution}}{M_{total}} \tag{8.7}$$

Figure 8.5 shows that the depth of the weathering front (Lwf) is driven by the amount of water (mean annual rainfall – evapotranspiration) and the amount and solubility of the mineral being considered (Eq. (8.7)). The slope of the concentration gradient $\left(\frac{dC}{dz}\right)$ is impacted by both the weathering front advance rate and the rate of mineral weathering. As weathering rate (R) increases or weathering front (ω) migration decreases, the slope of the gradient declines (and vice versa). The vertical distance over which soil mineral (or element) concentrations change from fully depleted to those of the parent material is the length over which the downward-migrating soil waters equilibrate (Leq).

One of the important questions in soil and water geochemistry is understanding the rate (R) of mineral weathering in natural environments. Mineral weathering rates can be determined in the laboratory, but there is general agreement that these do not always translate well to long-term natural soils. White et al. used Eq. (8.5) to determine mineral weathering rates for plagioclase in the Santa Cruz marine terraces. The term ω = (Lwf/soil age) and the slope of the weathering front (dC/dz) can be measured directly from data such as Figure 8.4c. The surface area of the mineral was calculated from the abundance of the mineral as determined by X-ray diffraction and a geometric model. From these observable data (and the determination of the soil age by cosmogenic nuclide analyses in Chapter 6), the rate of mineral weathering (mol m^{-2} mineral surface s^{-1}) can be calculated. White et al., using Na in the solid phase, found that plagioclase weathering rates in soils on the Santa Cruz terraces ranged from 4.2×10^{-16} to 1.37×10^{-15} mol m^{-2} s^{-1}.

Accurate measures of mineral weathering rates are important for understanding the role of soils in responding to, and moderating, climate via the consumption of CO_2[11] and for estimating rates of elemental release for ecosystems from soil processes. Biology has been implicitly included in the processes of chemical weathering (with its effects essentially embedded in some of the constants used earlier). To better understand the role of plants and biology, and to understand how weathering impacts the C cycle, one can describe chemical weathering of silicates as

$$2CO_2 + H_2O + CaSiO_3 \leftrightarrow Ca^{+2} + 2HCO_3^- + SiO_2$$

where the mineral $CaSiO_3$ is simply for illustration, but corresponds to the mineral wollastonite.

In this simplified reaction, the reaction of CO_2 and water creates the weak acid carbonic acid. The greater the CO_2 concentration in soil, the lower the equilibrium pH. As shown in Chapter 7, CO_2 concentrations in soil increase with biological production (and water and heat). An important part of this reaction is that the dissolution of a silicate releases a cation (Ca shown here, but also including Mg, Na, K, etc.) with bicarbonate (HCO_3^-) as the balancing anion. Thus, for every mole of silicate weathered (in this case), 2 moles of CO_2 are consumed. The dissolved Ca and HCO_3 have various fates. First, they may be transported in soil water to streams and rivers and eventually to the ocean. There, driven by biological biomineralization, the mineral $CaCO_3$ is formed and settles to the ocean floor:

$$Ca^{+2} + 2HCO_3^- = CaCO_3 + CO_2$$

The precipitation of the solid releases one of the CO_2 molecules originally consumed in the soil. This reaction can also occur in semiarid and arid soils (see later), where water from weathering is evaporated within the soil, accumulating the products of weathering close to their original source. Globally, the chemical weathering of silicates, which occurs to a large degree within soils, is now known to be a long-term stabilizing process on the global CO_2 cycle.[12] It is also temperature dependent, with rates increasing with increasing temperature.[13]

The examples we have just examined represent soil biogeochemical processes on one side of an important climatic boundary: where rainfall exceeds evapotranspiration, and thus water continuously (but slowly) moves through the soils, weathering minerals, creating new ones, and slowly removing soil mass. On the other side of this boundary, the drier side, water cannot remove mass, and soils tend to undergo accumulation of materials and volumetric expansion. This is examined in the next section.

8.3 Soil Biogeochemical Processes in Dry Environments

One of the important processes that impact all soils, even the humid ones considered earlier, is the slow deposition of dust, aerosols, and nutrients from the atmosphere (either as dry

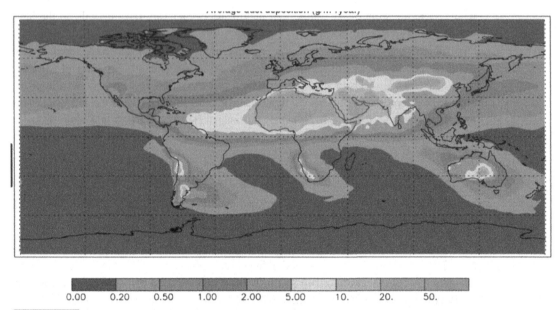

Figure 8.6   Estimated dust fluxes to land and oceans in g m^{-2} y^{-1}. From T. D. Jickells et al., Global iron connections between desert dust, ocean biogeochemistry, and climate, *Science*, 308: 67–71 (2005).

matter or in rainfall). It is particularly in arid environments that this process is magnified, both because there is a greater dust flux and because aqueous processes are less effective at removing the atmospheric inputs.

Dust is preferentially generated in dry environments, where plant cover is low and exposed sediment in dry stream channels, ancient lake beds, and other soft sedimentary outcrops provides ample material for wind deflation and transport. While much of the dust is redeposited locally (Figure 8.6), some is transported globally. For example, dust generated in northern and southern Africa is transported westerly across the Pacific and is an important source of elemental deposition in the Caribbean, southeast USA, and Brazil.[14] Dust generated from Mongolia is transported easterly across the Pacific and is an important source of nutrients, such as P, on the remote islands of the Hawaiian volcanic chain.[15] In humid regions, the dust may be rapidly incorporated into soils and landscapes, leaving little visual evidence of its importance (though mineralogical and isotopic investigations can detect its integrated effect). However, in arid landscapes, due to both the high deposition rates and the preservation of the deposition, the effects are magnified and represent an important, if not dominant, process in the soil biogeochemistry of these regions.

In arid landscapes, many depositional areas consist of gravelly alluvium. The slow deposition of dust and aerosols (silicate sand, silt, and clay; carbonates, sulfates, chlorides, nitrates, etc.) migrate into the coarse parent material matrix. The silicate dust in particular, due to its low solubility, concentrates near the surface and tends to loft coarse fragments upward, creating a gravelly "desert pavement" with an underlying horizon consisting of

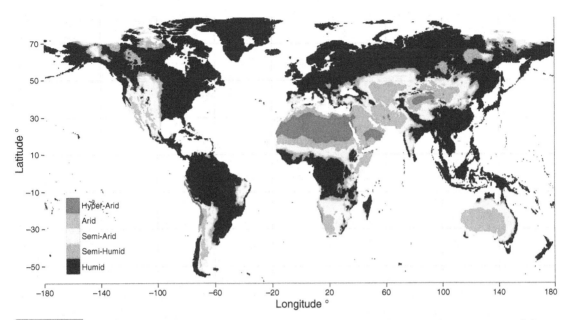

Figure 8.7   A map of the global distribution of climate zones. From http://ozewex.org/semi-arid-ecosystems-particularly-sensitive-to-mega-droughts/

dust and (if very dry) salts of various types.[16] The incoming salts migrate into the soil and accumulate.

Figure 8.7 shows the distribution of humid and nonhumid landscapes on Earth. The chemical weathering discussed earlier is typical of processes in the humid regions in the figure, which, in comparison to Figure 6.6, are also areas of low dust generation. The hydrological opposite of the humid climates is the hyperarid deserts on Earth: the Sahara, Namibian, Arabian, Mongolian, and Atacama deserts. Here, we focus on soil biogeochemistry in the Atacama desert, which is likely to the driest location on Earth, with annual rainfalls of ~1 mm y^{-1} in some locations and sometimes decades between rain events. As a result, the region is also largely abiotic (though some organic compounds, and occasionally active microorganisms, live in the soils).

Figure 8.8 shows the location of an ~2 My-old alluvial fan derived from past erosion on the hillslopes in the background. There are no plants, and the deeply inset truck tracks are indicative of the soft and porous dust and salt (anhydrite) layer that has accumulated at the land surface over time. The soil formed here, along with those from Santa Cruz, is likely one of the most geochemically well-studied soils on Earth.[17] Data for the soil are given in Table 8.2. Unlike the soils on the Santa Cruz terraces, where water is constantly moving downward, reacting with minerals, creating secondary minerals, and removing mass, the soil here is, from a first-order perspective, a passive collector of atmospheric deposition. The composition of atmospheric deposition in the Atacama Desert, for a period of a year, is given in Table 8.3.

Soil depth midpoint (cm)	Gravel (%)	Si	Al	Ca	Mg	K	Na	Fe	Ti
Table 8.2 Total chemical composition of the Yungay soil (shown in Figure 8.8)									
					Weight %				
1	17.1	30.48	8.24	2.98	0.91	1.03	2.71	2.26	0.29
2.5	0.7	9.63	2.71	17.41	0.38	0.34	2.50	0.85	0.11
7.5	11.9	18.63	5.01	10.45	0.49	0.64	3.55	1.16	0.15
19	24.9	16.83	4.42	12.10	0.51	0.63	3.70	1.21	0.18
32.5	24.6	16.84	3.94	12.60	0.59	0.72	3.94	1.08	0.15
55	10.0	22.08	5.51	10.10	0.82	0.86	5.11	1.50	0.20
86.5	10.6	24.79	6.13	8.03	1.09	1.02	5.26	1.96	0.27
112	0.4	21.87	5.45	6.48	0.88	0.97	9.53	2.20	0.27
134	0.0	5.64	1.36	1.19	0.17	0.25	34.74	0.47	0.06
150	0.0	16.91	4.43	3.47	0.65	0.64	14.28	1.94	0.21
177	0.3	25.84	6.25	5.89	0.85	1.07	2.02	2.82	0.32
186	4.1	28.30	6.25	4.90	0.67	1.24	2.00	2.30	0.26
202	1.9	25.76	6.23	6.21	0.86	1.03	1.96	2.84	0.33
222	0.0	25.01	6.77	5.67	1.08	1.02	1.96	3.18	0.37
	Bulk density (g cm^{-3})	Zr	Sr	Ba	P	S	Cl	Inorganic C	
		ppm			**Weight %**				
1	1.4	47.50	0.03	0.09	0.05	0.17	0.04	0.19	
2.5	0.8	22.50	0.01	0.04	0.03	14.34	0.01	0.21	
7.5	0.6	28.83	0.02	0.06	0.04	7.87	0.02	0.23	
19	1.2	28.67	0.02	0.06	0.05	9.18	0.03	0.20	
32.5	1.5	27.83	0.02	0.05	0.03	9.36	0.03	0.16	
55	1.3	37.33	0.04	0.08	0.06	7.35	0.66	0.15	
86.5	1.2	47.00	0.12	0.10	0.06	4.94	0.42	0.15	
112	1.5	63.50	0.07	0.09	0.05	2.79	5.73	0.39	
134	1.7	15.17	0.03	0.02	0.02	0.14	41.00	0.08	
150	1.7	189.49	0.03	0.06	0.04	0.86	19.03	0.39	
177	1.7	299.32	0.03	0.14	0.07	1.47	0.36	1.03	
186	1.6	354.98	0.03	0.14	0.06	1.03	0.19	0.92	
202	1.7	289.65	0.03	0.13	0.08	1.56	0.29	1.07	
222	1.7	236.32	0.03	0.11	0.09	1.57	0.32	0.88	

From Ewing et al., unpublished.

The deposition represents a mix of silicate minerals in dust (the Si) as well as the ubiquitous and biogeochemically important anions from aerosols (marine and recycled regional salt in the case of the Atacama Desert): Cl^- (from NaCl), NO_3^- (and smaller amounts of NH_4) largely from marine sources in the cold, upwelling waters in the nearby

Table 8.3 Atmospheric deposition chemistry for the Atacama Desert							
Site	Total (g m^{-2} y^{-1})	Si	Na$^+$	Ca^{+2}	Cl$^-$	NO$_3^-$	SO$_4^{-2}$
		mmol m^{-2} y^{-1}					
Yungay	4	22	20	6	4	17	19

From Ewing et al. (2006).

Figure 8.8   Photograph of Atacama Desert landscape and the soil (see excavation pile in forefront) discussed here. Note: the landscape has no plants. The imbedded wheel tracks reflect a surficial, low-bulk-density V horizon made of an admixture of silicate dust and CaSO$_4$.

Pacific, and SO$_4^{-2}$ from CaSO$_4$ from marine sources and from outcrops of ancient sulfate deposits in the desert.

While the present rate of deposition is not necessarily that which has occurred through vast periods of time, the general chemical makeup of the soils largely reflects a continued long-term input of this makeup to the landscape. As in other deserts, the annual inputs are at least partially retained by becoming lodged within and (after rains) under gravels at the surface. Additionally, fogs and dew help wet and then "cement" fine-grained particles together at the surface, reducing their susceptibility to subsequent deflation. Rainfall, as mentioned, is very rare but also highly stochastic and variable. There is geomorphic evidence for occasional past rainfalls that were one to maybe two orders of magnitude greater than the long-term

Table 8.4 A summary of key differences between the environment and soils at humid Santa Cruz, CA and the hyperarid Atacama Desert, Chile

Site	MAP (mm)	MAT (C)	Water flux (mm yr^{-1})	Water flux × age soil (m)	Volumetric change (cm/upper 200 cm)	Soil CO_2 (ppm)	pH	Na gain/ loss (kg m^{-2})	Si gain/ loss (kg m^{-2})
Santa Cruz	727	13.4	174	15 660	−312	5000–10 000	~6.7	−73	−1300
Atacama	1	16.4	0	~0	+95	~400	~7–8.3	480	−109

MAP, mean annual precipitation; MAT, mean annual temperature.

norm. Spread over millions of years, this dissolves the nitrates, chlorides, and sulfates and differentially transports them downward based on their relative solubility (Table 8.4). However, chemical weathering of the silicate minerals in the original alluvium is almost nil, and sand and gravel remain pristine in appearance after several million years. The soil profile thus expands (rather than contracts as in humid environments) and gains mass (rather than losing it) with the passage of time. Here, we use the methods introduced earlier to quantify these effects.

A profile sketch of the soil is provided in Figure 8.9. A quick review of horizon nomenclature is useful: V = horizon with vesicular porosity and low bulk density, y = sulfates, z = salts more soluble than gypsum (chlorides, nitrates here), n = sodium (NaCl here), m = indurated (impenetrable without power implements). Using data in Table 8.2, the volumetric expansion (epsilon) values can be calculated (Figure 6.10a). Virtually all horizons above 150 cm have undergone large expansions because of salt and dust accumulation. One can calculate the volume of expansion from this expression:

$$\Delta soilthickness = \left(\frac{V_s}{\varepsilon + 1}\right) - V_s \tag{8.8}$$

where V_s = thickness (or volume) of present soil layer.

By summing for all horizons, an approximate expansion, or uplift of the land surface, of 95 cm has occurred in the soil over the course of its development. The original fluvial sand and gravel that represented the original land surface have been lofted and incorporated into the now inflated upper 1 m of the soil and are diluted by atmospheric inputs. Using Ti as an index element, we can examine the tau values of S and Cl (which are externally derived from the atmosphere) and Si (which, of course, tends to be lost or translocated during chemical weathering reactions).

First, the Si tau values hover around + or − 0.1, variation expected from parent material and soil variability (Figure 8.10b). In contrast, S has an enormous tau value of 500 near the surface (accounting for the Vy2 horizon (Figure 8.9) and the large expansion in Figure 8.10). The tau values decline with depth as S decreases. It should be noted that

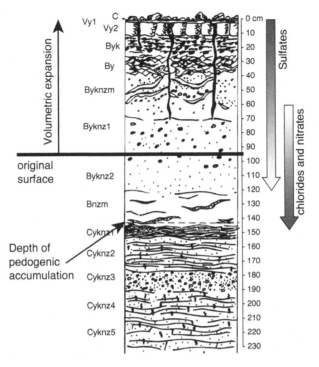

Sketch of the soil profile found in the landscape of Figure 8.8. Horizon nomenclature is shown on the left. Depth and trends in sulfate and chloride movement are illustrated on the right. The depth of significant pedogenic alteration is illustrated by the red dashed line at 145 cm, while the approximate original land surface is shown by the dark solid line at 95 cm. Sketch by S. A. Ewing.

sulfates dominate the chemistry and physical properties of the upper 75 cm of the soil. The soft, porous polygons of anhydrite (anhydrous form of $CaSO_4$) characterize the upper 10 cm or more and are seen in Figure 8.9. Second, dense and indurated polygons (with much large spacing between cracks) dominate from 20 to 70 cm (Figure 8.9). While much remains to be learned about the dynamics of sulfate cracking, it is likely that these are dynamic features on the timescale of centuries to millennia, and when rare large rains occur, the anhydrite forms gypsum ($CaSO_4\text{-}2H_2O$) and undergoes volumetric expansion. The long periods of drought likely cause the reverse process, driving crack and polygon formation. The large cracks also act as a conduit for physical movement of dust, salt, and preferential water flow in rare large rainfalls.

In contrast to S, the Cl tau values spike to over 3800 in the Bnzm horizon, accounting for its large volumetric expansion (Figure 8.10a,b). This horizon also contains a few weight percent of $NaNO_3$, an important salt from economic perspectives in this part of South America. While the vertical separation of very soluble NaCl and $NaNO_3$ from the less soluble $CaSO_4$ is intuitive, the conditions and events that form this are not the annual norm. Present average rainfall is incapable of more than wetting the land surface. To move, and separate from sulfate, the chlorides and nitrates to depths of more than 1 m requires considerable pulses of

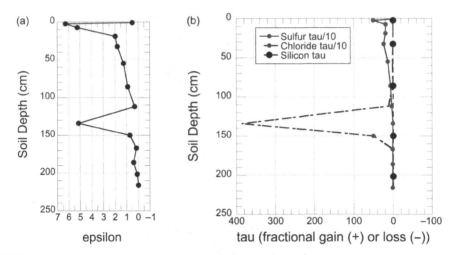

Figure 8.10 (a) Epsilon (expansion) and (b) tau values (fractional gains/losses) vs. depth for soil in Figure 8.9, using data from Table 8.2 and relationships given in this chapter.

rainfall, and multiple times, in order to accomplish the separation. On the other hand, these events can also not be so large or frequent as to simply dissolve and remove the soluble salts.

8.4 Integration

This chapter and others provide commonly used tools to employ soil analyses to better understand biogeochemical processes. Thus far, we have used the same principles to look at soils on distinct ends of a climate spectrum. Here, we assemble the data in a way that illustrates the first-order impact of climate on soil biogeochemical processes. A comparison of biogeochemical processes at Santa Cruz (humid) and the Atacama Desert (hyperarid) are listed in Table 8.4.

The largest climate difference between the two sites is rainfall. Over its lifespan, the California soil has experienced ~15,000 m of water passing through (and below) its upper 2 m, while the Chilean soil has experienced almost zero water flux through its profile. Additionally, rainfall drives biological activity, and the California soils commonly have 5,000 to 10,000 ppm of CO_2 in the soil atmosphere vs. essentially atmospheric values (400 ppm) in the Chilean soils. The California soils have lost up to 90 and 70 percent of the parent material Na and Si, respectively. In contrast, the Chilean soil has gained a large mass of Na and has lost an apparently smaller amount of Si (though this is probably within the error caused by parent material uncertainty, etc.).

This trend appears to exist on a larger scale. Amundson et al.[18] examined the tau values of the major elements in the upper 10 cm of soils in a range of rainfalls (Figure 8.11). The purpose was to determine whether soil surfaces systematically bear a fingerprint of rainfall in their chemical composition, one that could in turn be used to interpret the climate history from soil surface

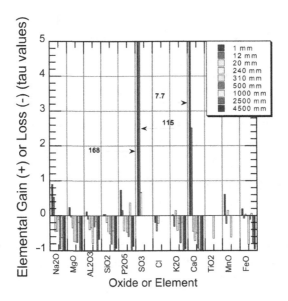

Figure 8.11 The gains or losses of major elements in the upper 10 cm of soils in differing climate regimes around the world. From Amundson et al., On the in situ aqueous alteration of soils on Mars, *Geochimica et Cosmochimica Acta*, 72: 3845–3864 (2008).

chemical analyses on Mars. The data accessed in that analysis reinforce what has been presented in this chapter, showing that the rainfall threshold, which marks the boundary between net losses and net gains of mass, is somewhere between 20 and 200 mm of MAP for many elements. We are not concerned here with the exact value of this threshold but rather, to gain a general appreciation of the changing impact of climate on soil biochemical processes.

8.5 Summary

This chapter took a field-scale view of ways to understand and quantify chemical weathering amounts and rates in soils. Mechanisms and models at a more molecular scale were ignored; instead, the focus was on tools applicable to field-scale data and problems. At the time of this writing, there are more than 530 papers citing the original 1987 paper by Brimhall and Dietrich[19] that reintroduced mass balance analyses to a more general soil audience, and the data in these papers can prove useful for regional and global comparisons for both educational and research purposes.

8.6 Activities

A detailed study of the *in situ* rates and mechanisms of soil biogeochemical weathering was conducted by White and colleagues on the marine terraces near Santa Cruz, CA. Total chemical analyses of soils, as well as soil pore water chemistry, were collected. Additionally, C, N, and other biological parameters were compiled.

Here, using the data from White et al. 2008, calculate and plot (vs depth) for each soil:

a. The Na tau values vs. depth
b. The total tau values (see later)
c. The epsilon (volumetric change) values
d. The total volumetric gain or loss for each profile

The "total tau" is the total mass loss or gain in the soil. It is similar to the other tau values except that it involves only the immobile element (here we use Zr):

$$tau_T = (Zr_{pm}/Zr_{soil}) - 1$$

Provide a paragraph that shows a narrative of how these soils change chemically and physically with time, and how the landscape itself changes with time.

e. Using the three youngest soils (due to clarity of the profiles), calculate the rate of the Na weathering front advance vs. time (use the depth at which you determine a large inflection in Na concentration from near zero values). Project the depth of this front to terraces 4 and 5, and comment on the nature of the relationship.
f. For the soil on terrace 2, the Na, Ca, and Mg content of the total soil, the fraction adsorbed by cation exchange capacity (CEC), and the concentration in soil pore waters are provided. Some questions include:

   i. Does the pore water become saturated at or near the Na and Ca weathering fronts?
  ii. The pore water is dominated by Na, and the exchange sites by Ca and Mg. Why do you think this might be the case?
 iii. Calculate the equilibrium partition coefficient for Na in solution vs. Ca, using the expression

$$K_{Na/Ca} = (C_{Na}/C_{Ca})_{exchange}/(C_{Na}/C_{Ca})_{pore\ water}$$

9   Soil Processes on Sloping Landscapes

It has long been recognized that soil properties vary with topography,[1] which has spawned considerable research on soil toposequences. While it is known that slope impacts the soil water balance, rates of weathering, and biological activity, soil research has long been hindered by an absence of a comprehensive physical model of hillslope processes and an appreciation of the rates at which soil erodes and is reformed by biophysical processes. To some degree, this has been the result of communication barriers between fields. As mentioned previously, some of the most significant conceptual advances of James Hutton were based on his intuition combined with observations of hillslope soils. He knew that soils on hillslopes occur virtually everywhere, that they appear to undergo constant erosion, and thus, that there must be regenerative processes operating to compensate for the soil loss. In the somewhat parallel universe of pedology, it is fair to say that many, though not all, soil toposequence studies tended to consider soils on slopes as quasi-static entities, whose properties hinged on slope-related differences in water balance and other factors due to their slope position. Hillslope soils are commonly called "residual soils," which are soils formed from, or resting on, consolidated rock of the same kind as that from which they were formed and in the same location.[2] This definition implies that these are not dynamic and do not have parent material that may have been derived from other locations upslope. However, this concept of hillslope soils is largely untrue and obscures the systematic nature of soil processes on hillslopes. Even in the most abiotic and driest areas of Earth, soils are in constant physical motion due to physical and/ or biological forces. The rapid embracement of this concept in the past 20 years is likely one of the most significant advances in soil formation theory in many decades and possibly the past 75 years.

If we attribute the recognition of the dynamic nature (and gain and loss) processes of hillslope soils to Hutton, it is to the American geologist G. K. Gilbert that we owe a more quantitative model for the processes. While working in the arid Henry Mountains of Utah, Gilbert contemplated the erosion of soil from convex hillslopes as well as the nature of the processes that convert the underlying rock into the overlying soil. This work[3] has informed and promoted research and modeling within the field of geomorphology ever since. However, it is only since the advent of cosmogenic nuclide chemistry applied to the geosciences that rapid and exciting advances in our understanding of hillslope soils have occurred. In this chapter, the basics of the advances in this field are examined, and what we now know about the dynamic nature of hillslope soils from the perspective of climate is reviewed in the context of biogeochemistry.

9.1 The Nature of Hillslopes

Most of the Earth's land surface is not level. Slope is a measure of the change in elevation of the land surface (z) with a change in horizontal distance (x), dz/dx. The first derivative of slope is curvature, which describes the change in slope with lateral distance. From this perspective, three general categories of land surfaces can be defined as a function of slope and curvature:

- Level (or stable) land surface: *slope (dz/dx)* = 0
- Erosional land surface: *curvature (d(slope)/dx)* = (−)
- Depositional land surface: *curvature* = (+)

Examples of these landscapes are illustrated in Figure 9.1. On the basis of curvature, erosional land surfaces are convex hillslopes, where slope increases in a downslope direction. Depositional land surfaces are concave, where the reverse occurs.

Soil on hillslopes is almost invariably a layer (or layers) of unconsolidated material, roughly a few centimeters to maybe 100 cm thick, over the underlying rock or sediment (Figure 9.2). An important step for pedologists examining hillslope soil profiles is differentiating between mobile soil and bedrock/saprolite. Mobile soil is the part of the profile where rock particles are no longer in place and do not retain their original rock fabric. This part of the soil has been biophysically lofted and mixed from the underlying rock material (Figure 9.2). Below this is a rock or saprolite contact. It is common, on many hillslopes, that the saprolite is indeed chemically weathered (and may contain accumulations of pedogenic clay, etc.), but it is also important to recognize that it still

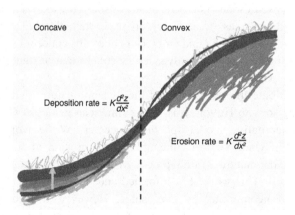

Figure 9.1   A simple hillslope cross section showing the convex and concave segments. The line represents a previous land surface, and the yellow arrow reveals the direction in which the land surface is moving over time. The illustration also shows that convex landscapes tend to have thinner soils than the concave portions, which accumulate sediment and organic materials over time.

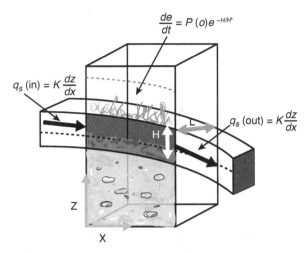

Figure 9.2 A representation of the mobile soil layer(s) on a convex hillslope. The flux in and out of the box (Volume per unit cross section L) is driven by slope. The downward wearing of the landscape (erosion) is the opposite sign of soil production, which replaces the removed material.

maintains much of its original rock structure. In the models discussed later, we are concerned only with the "mobile soil" layer mentioned here, which is commonly the "A" horizon(s) (or AB horizons) of the soil profile. In a general way, the mobile soil layer is a slow-moving conveyor of dark organic-rich soil material from convex portions of landscapes to adjacent hollows or other depositional areas. In the following, we examine how to express this numerically.

9.2 Soil Erosion

On a convex landscape, the rate of change of the soil thickness (mobile soil layer) (H), is:

$$\frac{dH}{dt} = Soil\ Production - Soil\ Erosion \tag{9.1}$$

The units of production and consumption are length (L)/time (T). On vegetated, convex hillslopes, erosion is a process driven by somewhat random biological soil particle movement and physical processes, which on average move in a net downslope direction. Examples of biophysical mixing are worm and gopher burrowing, tree throw, ants, freeze-thaw, etc. On a simple two-dimensional hillslope, this movement – which can be referred to as soil creep[4] – is represented as

$$q_s = -K\frac{dz}{dx} \tag{9.2}$$

where q_s = soil flux ($L^2\ T^{-1}$), K = transport coefficient ($L^2\ T^{-1}$).

Note that, while q_s commonly is reported in the units of $L^2 T^{-1}$, it can more intuitively be understood as volume of soil passing a given slope contour width ($L^3 L^{-1} T^{-1}$). It has also been proposed that soil creep can be dependent on soil thickness (H), which is

$$q_s = -KH\frac{dz}{dx} \tag{9.3}$$

where K has units of LT^{-1}.[5]

This model is reasonable because the soil thickness also impacts biological processes and movement. For a given soil box (Figure 9.2), the net erosional flux is

$$\frac{dq_s}{dx} = -K\frac{d^2z}{dx^2} \tag{9.4}$$

where $\frac{d^2z}{dx^2}$ is the curvature. Thus, by knowing the curvature (and the value of the local transport coefficient), one can calculate rates of soil erosion at a site. Later, we will examine some ways that K is estimated (see also Box 9.1).

Box 9.1   **First Measurement of Slope-Dependent Soil Creep**

While G. K. Gilbert deserves and receives considerable recognition for his research on hillslopes and the nature of the processes that shape them, a less recognized contributor is Charles Darwin. While Darwin's final book on earthworms is gaining more recognition from geographers and soil scientists for the importance of biology and mixing during soil formation, what has not been recognized is that Darwin, in that book, was likely the first person to formalize and quantify the rates of slope-dependent soil creep, in his case caused by earthworms. He wrote:

> It was shown in the third chapter that on Leith Hill Common, dry earth weighing at least 7.453 lbs. was brought up by worms to the surface on a square yard in the course of a year. If a square yard be drawn on a hill-side with two of its sides horizontal, then it is clear that only 1/36 part of the earth brought up on that square yard would be near enough to its lower side to cross it, supposing the displacement of the earth to be through one inch. But it appears that only 1/3 of the earth brought up can be considered to flow downwards; hence 1/3 of 1/36 or 1/108 of 7.453 lbs. will cross the lower side of our square yard in a year. Now 1/108 of 7.453 lbs is 1.1 oz. Therefore, 1.1 oz of dry earth will annually cross each linear yard running horizontally along a slope having the above inclination (9° 26′); or very nearly 7 lbs will annually cross a horizontal line, 100 yards in length, on a hill-side having this inclination.

If we focus on the last line, assume the soil has a bulk density of 1.5 g cm^{-3}, and convert slope and length to the appropriate units, we can derive a soil transport constant of about 126 cm^2 y^{-1}. Later, when we review seminal research in the coast range of California, the authors there found values of 50 cm^2 y^{-1}. In England, the soil creep was driven by earthworm casts thrown up on the land surface (and the collapse of the burrows, as Darwin astutely noted). In California, creep is driven largely by gopher burrowing. Thus, the larger K value for England is consistent with the more humid climate and the rapid rates of soil movement by worms that Darwin famously quantified in his book.

9.3 Soil Production

It is likely that many largely undisturbed hillslopes approach a steady-state condition where soil production = net erosion, and thus at a given location soil thickness is relatively constant.

The rates, and the behavior, of soil production have been speculated on since G. K. Gilbert hypothesized about this in the late nineteenth and early twentieth centuries.[6] Briefly, he suggested that because the effectiveness of both biological and abiotic processes decreases with increasing soil depth, the rate at which they can dislodge particles and create new mobile soil should decline with increasing soil thickness. This means that soil production rates in a given location are lower on thicker soils than on thinner soils. An additional long-considered question is how soil production behaves as the soil becomes exceedingly thin. One possibility is that as soil reaches some critical minimum thickness, biology can no longer thrive, and thus rates of production begin to plummet (producing what is a called a hump-shaped production function (Figure 9.3)). Alternatively, rates of production may increase smoothly with decreasing thickness, as illustrated in Figure 9.3. These will be discussed further.

Empirically determining the rates of soil production on hillslopes was largely not feasible for more than a century. A fundamental breakthrough occurred when the use of cosmic isotope geochemistry was applied to the problem.[7] Heimsath and his coauthors used cosmogenic ray production of ^{26}Al and ^{10}Be in quartz to calculate the rate of soil production for soils of differing thicknesses on gentle soil-mantled hillslopes in coastal California at Tennessee Valley. Briefly, the downward erosion rate (ε), of opposite sign to soil production, is

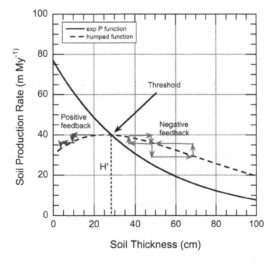

Figure 9.3 An example of an exponential soil production function (from Tennessee Valley, CA) and a humped soil production function. As discussed later, for the humped production function, the maximum production at H = 25 cm is a threshold between two alternate states of the soil system.

$$\epsilon = \frac{\Lambda}{\rho_r} \left(\frac{P_{(h,\theta)}}{C} \right) \tag{9.5}$$

where Λ = the mean attenuation length of cosmic rays in soil ($\sim$165 g cm^{-2}), ρ_r = parent material bulk density, $P_{(h,\theta)}$ = the nuclide production rate at depth h and slope θ, and C = concentration of the radionuclide in the mineral quartz (atoms g^{-1}).

The resulting data suggested a soil production function of the form

$$P(H) = P(0)e^{-Hm} \tag{9.6}$$

where $P(H)$ = production at a given soil thickness (common units, in terms of length, are m My^{-1}), $P(0)$ is the production at depth 0 (for a nonhumped production function), H = soil thickness, and m is a constant.

Figure 9.3 shows the data for the original 1997 paper for coastal California. Soil production rates increased from about 20 to 77 m My^{-1} with decreasing soil thickness – providing numerical support to Gilbert's hypothesis about depth-dependent rates of soil production. In the final chapter, the results of hundreds of other analyses from around the world will be compared, which provides implications for the sustainable use of hillslope soils.

9.4 Steady-State Hillslopes

Equation (9.1), at steady state, can be expressed as

$$\frac{dH}{dt} = 0 = P(0)e^{-Hm} - K\frac{d^2z}{dx^2} \tag{9.7}$$

where production rates are balanced by net erosion rates. By knowing the local production function, the relationship can be modified to determine the value of K.

9.5 Relevance to Soil Biogeochemistry

The realization that soils on hillslopes are in slow but constant motion, and are part of the processes that shape the Earth's surface, enables one to begin to put these rates into a framework that allows us to understand the implications for Earth surface processes.

9.5.1 Variation in Rates of Soil Production

From the preceding introduction, it is known that production rates of soil tend to decrease exponentially with soil thickness. However, how does the P(0) rate change in response to climate? For example, increasing rainfall and temperature should increase rates of biological processes and chemical weathering, enhancing the rate of landscape lowering and,

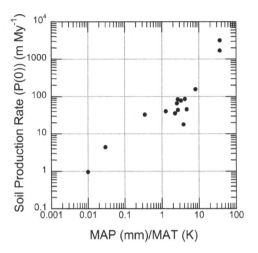

Figure 9.4 A compilation of soil production functions from a vast array of climate conditions, showing the variation in P(0) rate with increases in effective moisture (MAP/MAT). Data from Amundson et al., Hillslope soils and vegetation, *Geomorphology*, 234: 122–132 (2015).

hence, soil production. Recently, due to growing data availability, some global-scale comparisons have been made, showing indeed that the production rates respond to climate.[8] Here, one study is examined in which P(0) values were plotted against mean annual precipitation (mm) divided by mean annual temperature (°C), a metric that broadly reflects effective rainfall (Figure 9.4). As the data show, *P(0)* increases with mean annual precipitation (MAP)/mean annual temperature (MAT) or with MAP independently.[9] The rates increase from ~1 m My^{-1} in the hyperarid Atacama Desert to >2000 m My^{-1} in the high-rainfall mountains of New Zealand.

What do these rates reveal about soil-related features? First, residence time is considered. As was discussed earlier in the book, the age of soils on dynamic landscapes requires different concepts than for soil on stable land surfaces. Figure 9.3 illustrates that soil enters from above (through creep) and below (through soil production) and is lost via creep. One approach of assigning time is through residence time, or the time it takes for the thickness of soil to be replaced by new material from the underlying rock or saprolite:

$$\text{Residence time } (\tau) \ = \frac{H}{P(0)e^{-Hm}} = \frac{H}{K\frac{d^2z}{dx^2}} \tag{9.8}$$

One might note that this is not the only metric for residence time of a soil box. As Figure 9.2 illustrates, a second definition (leading to a much shorter residence time value) would be to also include the rate of incoming soil from creep upslope. This has been explored recently,[10] but given the more complex calculations involved, here the focus is on the commonly used metric in Eq. (9.8). For the California coastal range soils in Figure 9.3, it is observed that residence times of the soil vary from 0.1 to 130 Ky depending on soil thickness (Figure 9.5). Globally, assuming a soil of 50 cm, the residence times vary from 527 to 2 Ky, with τ decreasing with increasing effective rainfall.

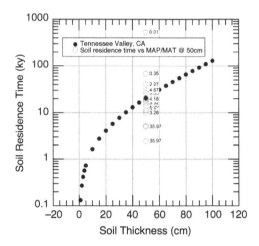

Figure 9.5   A calculation of soil residence times vs. soil age for Tennessee Valley, CA, and for soils in a broad spectrum of climate conditions (in Figure 9.4) assuming a thickness of 50 cm.

Since hillslope soils can be viewed as slowly moving conveyor belts of soil mass from uplands, one might also consider how long it takes for soil to creep off a hillslope. To estimate this rate, the following relationship is useful:

$$\text{Velocity (L/T)} = \frac{K}{H}\frac{dz}{dx} \qquad (9.9)$$

As an example, for the soils in the California coast, $K = 50$ cm^2 y^{-1}, the average soil thickness is around 50 cm, and an intermediate slope value is around 0.27. Using these values and Eq. (9.9), the time for soil (a sandy clay loam) to migrate 100 m is on the order of 37 000 y. In contrast, at a nearby location where the soil is formed from shale and is a clay-rich Vertic soil, the value of K is much higher (360 cm^2 y).[11] Here, the approximate travel time for 100 m is only about 5000 y. These back-of-the-envelope estimates are useful to gain an appreciation of the pace of soil and landscape processes.

The residence time of the soil, and its biogeochemical importance, was anticipated by James Hutton several hundred years ago. He speculated that erosion and replacement of soil is important to maintain a supply of nutrients for plants. We now know enough about rates of processes of important plant elements to test this hypothesis. Two key plant nutrients are P and N. P is derived largely from apatite in the soil parent material, while N is largely derived from atmospheric inputs and buildup over time. Long-term chemical weathering slowly depletes soil of P even though biological cycling works to retain this important element. Thus, old soils tend to be depleted in P or have their P bound with organic matter and secondary minerals in sometimes relatively unavailable forms. In contrast, it requires 10^2 to 10^3 y for soils to achieve N steady state, and thus, very young soils may be relatively impoverished in N. There are now enough quantitative soil data to begin to query how physical processes on hillslopes regulate plant nutrient availability.

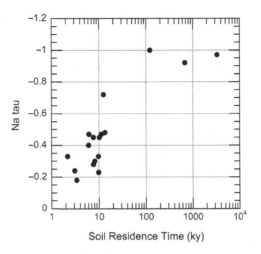

Soil Residence Time (ky)

Figure 9.6   The fractional loss of Na from granitic soils in North America vs. their residence time. The long-residence-time soils are found in the southeastern USA and Puerto Rico, where high rainfall and temperature create thick soils and accelerated losses of nutrients. Data from Rasmussen et al., Strong climate and tectonic control on plagioclase weathering in granitic terrain, *Earth and Planetary Science Letters*, 301: 521–530 (2011).

While the data for total P in hillslope soils are limited, there are ample data for total Na in soils and the associated soil production rates. Na is abundant in the silicate mineral albite, and albite has roughly the same weathering rate as apatite (the P source). Thus, we here use Na as a proxy for P since P data are not reported for all these sites. An important caveat is that P is less soluble than Na and may be more heavily biocycled than Na. Data from multiple sources[12] show that as hillslope soil residence time increases, the fractional loss of Na also increases (Figure 9.6). The soils with the very long residence times (10^5 y) are found in the humid tropical and subtropical forests. At shorter residence times, however, typical of less tropical conditions, soils maintain weatherable Na (and P). In terms of N availability, total N (and C) is correlated with soil residence time, though it is important to note that one reason why there is low N in rapidly cycling soils is because they tend to be shallow and have a lower volume for total accumulations of organic nutrients.

There are complexities to this model. In one example, for the humid basaltic soils of Hawaii, Porder et al.[13] have found that the sloping landscapes of the area have adequate P, while the level and slowly eroding landscapes are nutrient poor. In contrast, Eger et al.[14] found that, in both California and New Zealand, the rate of erosion was not strongly linked to P content (or other P metrics), because the rock being converted into soil has already undergone significant chemical weathering before becoming part of the mobile soil layer.

Thus, at this stage of our understanding, the rates of the biophysical soil processes of erosion and production are found to be dependent on both soil thickness and climate. Because of the restricted ranges in soil residence times, most hillslope soils might be hypothesized to lie in a residence time "sweet spot" where they are both N and P sufficient, but this is an excellent topic for much further research. Next, we examine more deeply some complex patterns that may underlie hillslope soils.

9.5.2 Hillslope Soils as Complex Systems

While a casual glance at a soil-mantled hillslope may reveal little to initially marvel at in terms of complex process behavior, the processes covered here underscore that these soils are complex systems. In complex systems, the various components of the system can interact with each other and can have properties or behavior that arise from interactions such as feedback loops. In other words, the system has ways of regulating itself and returning to original conditions if perturbed. Another feature of complex systems is that they may have thresholds above or below which the system might fundamentally, and permanently, change its state. Here, we examine these questions, which have begun to be explored conceptually.[15]

In feedback loops, the nature of the system itself impacts the rates of the processes that affect it. A simple cause–effect diagram of hillslope soil thickness and the processes that impact it is shown in Figure 9.7. As Eq. (9.6) indicates, as soil thickness increases, production decreases (dotted line in Figure 9.7), or *vice versa* as soil thickness declines. In its entirety, the inverse correlation of thickness with production, and the positive correlation of production with thickness, constitutes a *negative feedback loop*. A negative feedback loop has the effect of acting as a governor or a regulator on a system property, decreasing the amplification of a disturbance and returning the system to normal.

This is illustrated in Figure 9.3, which shows the two possible soil production functions. For the exponential function (considered here), any change in the thickness drives a corresponding change in production that drives the soil back to its original state (assuming that erosion rates remain constant). The possible humped production function offers an additional feature of soil complexity. In this case, there is a critical soil thickness value (H'). For soil thicknesses >H', a change in thickness ultimately will change production rates and move the soil back to its original state. However, if the soil thickness is reduced below H', it passes through a *threshold* and is subsequently driven to a soil thickness of 0 (see Figure 9.3). If soil production functions indeed have this behavior, the landscape would be expected to be a bimodal pattern of soil-mantled areas and bare rock. There is considerable work being done on these questions and on which models best describe a given location, but these potential endmembers reveal much of the complexity that underlies the production side of the soil mass balance.

Figure 9.7   A simple cause–effect diagram of the relationship between hillslope soil thickness and the two processes that control it. Dashed lines indicate an inverse relationship while solid lines indicate a positive relationship. The combination of a positive and a negative relationship yields a negative feedback loop.

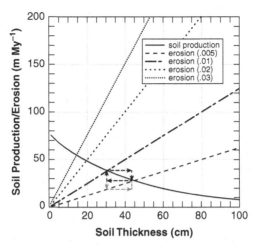

Figure 9.8 Empirically determined relationships between soil thickness and soil production rates and soil erosion rates for Tennessee Valley, CA. Erosion rates are calculated for four different values of curvature (m^{-1}) and Eq. (9.3) (following A. Heimsath et al., The illusion of diffusion: field evidence for depth-dependent sediment transport, *Geology*, 33: 949–952 (2005)). The dashed arrows reveal how both soil production and soil erosion respond to a hypothetical event that thins the soil. The soil thinning leads to increased rates of production, while the thinning leads to decreased rates of erosion, both of which drive soil thickness back to its original condition.

As with production, there are multiple candidates for models of soil erosion (see Eqs. (9.2) and (9.3)). If model 9.3 applies, there is also another negative feedback loop between soil thickness and (in this case) soil erosion. This creates a strong set of entwined negative feedback loops on soil thickness (Figure 9.8). This unique set of feedback loops suggests the tendency, if climatic conditions allow the presence of plants,[16] for hillslopes to maintain a soil cover. These possible feedbacks provide a mechanism for Gilbert's observation: "Over nearly the whole of the earth's surface there is a soil."[17]

9.6 Depositional Landforms

The discussion of hillslopes thus far has focused on the convex, erosional portions of the landscape and the soil biogeochemical processes that occur there. Invariably, net erosive landscapes merge with depositional settings. Here, we begin by considering concave land-scapes adjacent to convex hillslopes (Figure 9.1). In depositional settings, the rates of soil production can be viewed as being negligible, so that the change in soil thickness is the same as the deposition rate, which is the net deposition of sediment via soil creep:

$$\frac{dH}{dt} = -K\frac{d^2z}{dx^2} \qquad (9.10)$$

Soils in depositional settings tend to be thick and have significant quantities of C and N. Soil classification systems use the term "cumulic" to denote soils in depositional areas that accumulate sediment and C to create thick A horizons. Here we are focusing on undisturbed, vegetated landscapes as opposed to agricultural fields, where the mechanisms and rates of erosion differ greatly from those introduced thus far.

The importance of ongoing erosion and deposition of soil on hillslopes is important to soil properties, such as organic C and N, only if the rates of erosion become large relative to the rates of the soil processes of interest. For example, depositional areas in soil-mantled landscapes commonly have higher total C storage (in terms of mass per unit area) than erosive portions of the landscape (Figure 9.9c). Yoo et al.[18] examined the hypothesis that the erosion of soil C from convex portions of slopes (Figure 9.9a) is a reason for the large amounts of C found in depositional segments (Figure 9.9c). However, after quantifying rates of erosive C loss in comparison to rates of net primary production (NPP) and rates of organic matter decomposition, they found for Tennessee Valley that most eroded C is lost via biological cycling by the time the soil reaches a depositional setting. For this landscape, the total C in soil is related to soil thickness, and the deep and thickening soils in depositional areas are largely accumulating *in situ* produced organic matter from plant production. The research on the linkage of soil biophysical transport and soil biogeochemistry is still in its infancy, but the work with soil C provides a template for considering other soil properties as well as the nature of the processes in other climates and lithologies.

9.7 Summary

The ability to qualitatively and quantitatively study the nature and processes of soils on hillslopes is an exciting development in recent Earth science. Like many areas of soil biogeochemistry, it involves the wedding and interaction of different fields of science and the linkage of ideas across hundreds of years of time. The paper by Heimsath et al. (in 1997) that introduced the use of cosmogenic isotopes to constrain the soil production function in several locations considered in this chapter has over 500 citations at the time of this writing. In addition, thousands of analyses by many research programs are adding to this database. As a result, we can now look at upland landscapes and view them through a powerful conceptual lens, a lens that allows us to perceive their dynamic, and complex, behavior. We can now add numerical values to concepts that Hutton and Gilbert so astutely generated centuries ago.

9.8 Activities

Field scientists can conduct surveys or use lidar data (or, at a coarser scale, satellite data) to generate digital elevation models (DEMs) of hillslopes. Then, using software (for example,

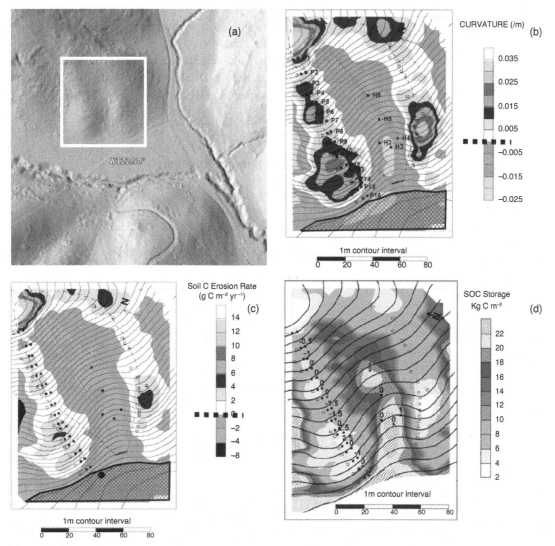

Figure 9.9   (a) A Lidar-based shaded relief image of two hillslopes in Tennessee Valley, CA. Area in white box is illustrated in the following figures. (b) Negative curvature, where positive values are convex landscapes and negative values are concave and depositional. (c) Soil C storage for the entire soil profile as a function of landscape position. (d) Rates of soil C erosion (which are driven by curvature). Images (b), (c), and (d) from Yoo et al., Erosion of upland hillslope soil organic carbon: coupling field measurements with a sediment transport model, *Global Biogeochemical Cycles*, 19: GB3003 (2005).

the freeware QGIS), slope and curvature can be calculated on the geospatial template. Slope is d(elevation)/d(horizontal distance) and curvature is the first derivative of slope: d(slope)/d(horizontal distance). These calculations take into account the irregularity of a slope in multiple directions.

In this book, for simplicity, we examine processes in just two dimensions (vertical and horizontal), essentially assuming that the slopes we are examining do not change laterally (for example, the escarpment of a stream terrace).

For a typical soil-mantled hillslope in central coastal California (Tennessee Valley) the elevation/distance profile was extracted from a DEM, and a second-order polynomial model was fitted to describe it (total distance from crest = 0 to base = 230 m):

$$\text{Elevation (m)} = 143.3 + -0.0646(x) + -0.0079(x^2) + 3.5366 \times 10^{-5}(x^3) + -4.0762 \times 10^{-8}(x^4)$$

9.8.1 Determine Equations

Determine the equation for the slope vs. horizontal distance and curvature vs. horizontal distance. A strong relationship exists between curvature (m^{-1}) and soil thickness:

$$\text{Soil thickness (cm)} = 80 + (80/0.04)(\text{curvature})$$

9.8.2 Create Plots

Create plots of:

  i)   The hillslope elevation profile
  ii) The hillslope slope vs. distance profile
  iii) The hillslope curvature vs. distance profile
  iv) Soil thickness vs. distance profile

9.8.3 Identify Landscape Segments

Using the most relevant metric to identify the convex vs. concave portions of the slope transect, identify the convex and concave landscape portions.

The rate of soil production ($m\backslash My^{-1}$) = $77e^{-023*T}$, where T = soil thickness (cm)

9.8.4 Residence Time

For the convex portions of the landscape (for which the soil production equation applies), plot soil residence time vs. horizontal distance.

9.8.5 Rates of Erosion

For two locations (50 m and 190 m; convex and concave) calculate the rates of soil erosion, assuming a K value of 50 $cm^2 y^{-1}$.

9.8.6 Impacts on Chemistry

From questions in Chapter 8, we know the rate of the Na depletion front for soils on level land in central coastal CA. Would you expect (and why) Na to be depleted, or not depleted, on the hillslopes examined here?

9.8.7 Impacts on Biogeochemistry

The N deposition rate in coastal CA is about 10 kg N ha^{-1} y^{-1}. The soil parent material has negligible N. For the convex soil in Section 9.8.4, if the average C% is 1.5 percent and the C:N ratio is 15:1, how does the rate of N deposition per m^2 compare with rates of erosion of soil N (assuming a bulk density of 1.4 g cm^{-3})?

Humans and Soil Biogeochemistry

We live on a used planet,[1] where about half of the Earth's land surface is under our direct management and the remainder is subject to our influences through other impacts (Figure 10.1). Farming, combined with the acquisition and use of fossil fuels, is now impacting soils directly through farming and indirectly through the changing climate system. In this final chapter, the focus is on the human impact on the biogeochemistry of soils and its significance to society in this century.

This chapter is not a compendium or literature review of what we know about human impacts on soils but, rather, an introduction to the magnitude of the issue, how one can ask the questions needed to understand human impacts, and how to obtain data and make calculations that address the questions. This has been the principle used in other chapters in this book as well.

10.1 Natural vs. Domesticated Soils

Soils that have been converted to farmland or rangeland bear the properties that they acquired during their long evolution during soil formation, but they have also acquired new features and processes as a result of their use. In an analogy to Darwin's distinction between natural and domesticated animals, human-modified soils have been called *domesticated soils* as a way to recognize our profound impact on them.[2] Domesticated soils share a "genetic" linkage (in other words, some biological, physical, and chemical similarities) to their undisturbed counterparts but have acquired properties – or lost properties – as a result of their long interaction with humans. Based on the methods by which we classify soils and the detail of soil mapping, there are more than 20,000 soil series in the USA.[3] If this soil/area relationship is extrapolated to the globe, it is likely that there are a few hundred thousand different soil series on Earth. Of course, as discussed earlier in the book, the soil mantle is a continuum, and this distinction of individual soil types is a human construct that allows us to understand and quantify the immense range of soils.

Soil types, like biological species, are found in varying abundances – some widely spread and some quite local. When the footprint of farming and urbanization is laid over this map of soil diversity, it becomes apparent that some natural soil types have become exceedingly rare or in some cases, essentially "extinct."[4] Does this matter? In this chapter, this question is explored and its significance to global society is considered.

As an approximate starting point, undisturbed soils are largely near a steady state for many properties (at least on very short-term timescales) (Figure 10.2): soil C is added at

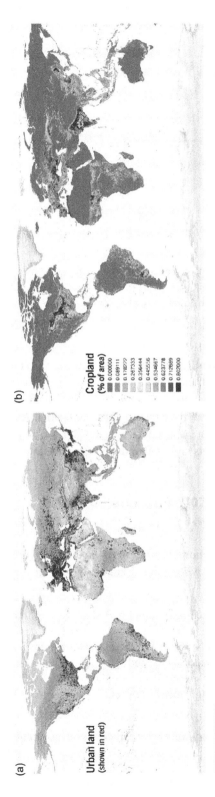

(a)

Urban land
(shown in red)

(b)

Cropland
(% of area)
- 0.000000
- 0.089111
- 0.178222
- 0.267333
- 0.356444
- 0.445556
- 0.534667
- 0.623778
- 0.712889
- 0.802000

Figure 10.1 An illustration of the physical human footprint on the global soil mantle: (a) locations of urbanized land and (b) areas of cultivated cropland. From R. Amundson et al., Soil and human security in the 21st century, *Science*, 348: 1261071 (2015). doi:10.1126/science.1261071

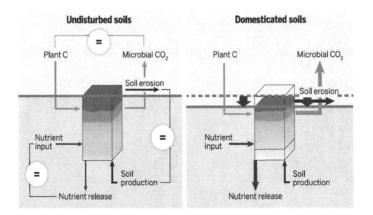

Figure 10.2 A conceptual analysis of how key processes in undisturbed soils and their domesticated counterparts operate. The hypothesis is that undisturbed soils are largely approaching steady state, while the impacts of agriculture, climate change, and other effects lead to imbalances in important soil processes. From R. Amundson et al., Soil and human security in the 21st century, Science 348, 1261071 (2015). doi:10.1126/science.1261071

roughly the rate it is lost by microbial respiration, soil production occurs at a rate that roughly matches soil erosion, and nutrient uptake by plants roughly matches the rate at which it is returned. Farming, urbanization, and secondarily climate change disrupt these approximate balances, which in turn can have tremendous implications for society. Thus, domesticated soils are commonly not at steady state and are far from their original condition. The magnitude of these shifts is considered next for some important properties.

10.2 The History of Human Soil Use

In order to attain a perspective of the enormity of farming on the Earth's surface, consider the following synopsis of this activity, written in a dramatic form here to help us consider farming with a fresh perspective:

> The last major global cataclysm occurred within a period of just a few thousand years: 20 percent of the planet's surface was physically scoured, massive quantities of sediment were produced, the global C and N cycles and the planet's hydrology were profoundly altered, which in turn caused climate change and wholesale changes in the Earth's flora and fauna. This upheaval was not the consequence of the Last Glacial Maximum (LGM), or the ecological changes following a meteor impact, but the results of the expansion of farming.

This statement may seem extreme, or entirely incorrect, if one has spent any time in the subdued and calming landscapes of rural agricultural America: there are few people,

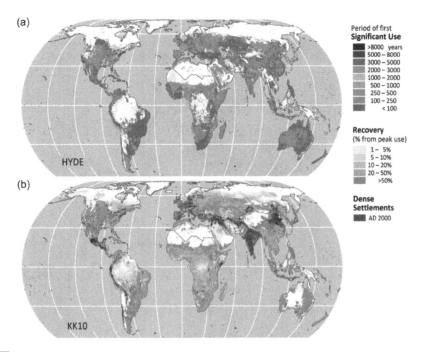

Figure 10.3   Two model-based estimates of the history of human land use during the Holocene. From E. C. Ellis et al., Used planet: A global history, *Proceedings of the National Academy of Sciences*, 110: 7978–7985 (2013).

generally very low traffic levels (except on the coasts), and miles of pleasing, and productive, agricultural mosaics. Yet, this misperception of what appears to be an idyllic landscape is a result of what some call "environmental amnesia,"[5] the fact that present generations lack a clear appreciation of the recent past, particularly the environmental changes from one generation to another. Farmlands are some of the most drastically altered ecosystems on the planet. In many locations, such as the Great Plains of North America, one is pressed to find even tiny remnants of undisturbed soil, let alone intact flora and fauna. This human alteration of the planetary norm began 10 000 years ago and has had a changing footprint as some lands have become less desirable due to degradation and as climate has gradually shifted (Figure 10.3).

Amundson et al.[6] examined the number and abundance of soils in the USA and the impact of agriculture and urbanization on this diversity (Figure 10.4a) As might be expected, not all soil series in the USA have equal distribution, and some are naturally of low abundance and regionally restricted. Thus, the authors defined *rare* soils as those with naturally <1000 ha total area and *unique* (endemic) soils as those that exist only in one state. If human use eliminated more than half the area of the natural expanses of these two categories, the soils were considered *endangered*, and if >90 percent was under intensive use, the soils were considered to be *extinct*. The distribution of endangered soil series in the USA is shown in Figure 10.4b. Some highly productive agricultural states (e.g. Indiana and Iowa) have 80 percent or more of their naturally rare soils in an endangered state.

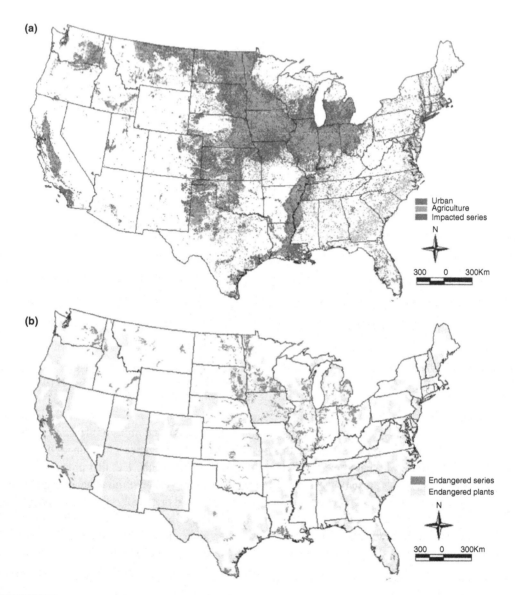

Figure 10.4 (a) The distribution of agricultural and urban land in the USA and distribution of soil series that have lost >50 percent of their original extent to these activities. (b) The distribution of endangered soils, and counties with endangered plants, in the USA. From R. Amundson et al., Soil diversity and landuse in the United States, *Ecosystems*, 6: 470–482 (2003).

As the authors noted, domesticated soils (Figure 10.2b) have enormous societal importance, yet preserving remaining tracts of undisturbed soils has potential values that were originally identified as reasons for saving global biodiversity:[7] (1) aesthetic qualities, (2) ecosystem services, (3) economic and biotechnology value due to biodiversity, (4) ethical reasons, and (5) science and education importance. Here, with the focus on biogeochemistry, the value of ecosystem services (carbon cycling), economic and biotechnology (micro-biodiversity), and science and education is examined. The importance of natural ecosystems as benchmarks to quantify the human impact on landscapes is essential for environmental research. In Chapter 1, the importance of the availability of natural soil systems along gradients of space and time underlies the Critical Zone Observatory framework. The value of natural soilscapes as benchmarks for science and society is a topic deserving of further discussion.

We begin an assessment of the impact of humans[8] on soil properties by focusing on farming, examining the organic and inorganic impacts we have had, and continue to have, on our planetary surface.

10.2.1 The Effect of Farming on Soil Carbon

One of the first innovations in the technical revolution that involved the domestication of species was the invention of tillage and cultivation. The removal of the native flora, combined with physical disturbance of the soil, accelerates soil microbial decomposition of organic C and N. As discussed earlier, much remains to be understood about soil C and N pools and what controls them, but there is evidence that the physical disruption of aggregates allows microbes to access C compounds and enhance their rates of metabolism.

As discussed earlier, the soil organic carbon (SOC) balance in an undisturbed soil is

$$\frac{dC}{dt} = I - kC \tag{10.1}$$

The solution to this equation (and with no C at t=0) is

$$C = \frac{1}{k}(I - Ie^{-kt}) \tag{10.2}$$

There are now several studies and data compilations of SOC (and N) before and after cultivation.[9] In general, at very low initial SOC contents, farming tends to increase SOC (due to higher plant inputs as a result of irrigation in what were likely initially arid conditions), while as the original C content increases, the amount lost tends to increase (Figure 10.4). This does not provide process or rate information. To do so, one can take advantage of "anthroposequences,"[10] soils under cultivation (but with similar state factors) for differing periods of time. Unfortunately, studies of anthroposequences are rare. Here, data from Jenny's study of the effect of farming in Missouri are used,[11] assuming that C loss is proportional to N, the element that Jenny measured.

The loss of C is due to an imbalance in inputs and outputs. If the initial C ($C_{t=0}$) in a soil is considered to be at steady state, time-dependent changes due to agriculture and tillage can be described as

$$C = \left(C_{t=0} - \frac{I_f}{k_f}\right)e^{-k_f t} + \frac{I_f}{k_f} \tag{10.3}$$

where k_f = decomposition rate under farming and I_f is plant C input under farming.

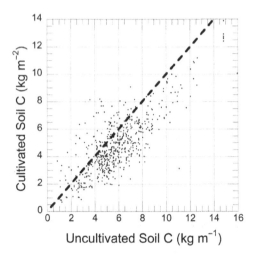

A comparison of C in uncultivated vs. cultivated soils. Data from L. K. Mann, Soil carbon storage after cultivation, *Soil Science*, 142: 279–288 (1986).

This can be rearranged to

$$C = (C_{t=0} - C_{ss})e^{-k_f t} + C_{ss} \tag{10.4}$$

where $C_{ss} = I_f/k_f$, or the steady-state value of the soil C pool after some decades of farming.

In Figure 10.6, a model fit of Eq. (10.4) is compared with the data of Jenny. Curve fitting can derive the value of k_f, in this case $0.017 \, y^{-1}$. From this value, and the projected steady-state endmember, one can calculate the inputs into the system as well ($k_f C_{ss} = I_f$). A first derivative of the exponential model fit to the data shows that the rate of C loss is greatest during the first years of farming and then declines with time.

The loss of SOC (and in rough proportion, N) – due to both increased values of k and sometimes decreasing values of I – has significance for food production and climate. First, until the invention of the Haber–Bosch process, which uses energy to convert atmospheric N_2 to NH_3, crop production in much of the world hinged on N availability in soils. Initially tilling soils caused rapid C mineralization and the release of N, but as the figures earlier indicate, a new and much lower steady-state condition was rapidly reached within decades. Thus, until after World War II, when industrial fertilizers became more commonly used, the yield per acre of major crops in the USA did not change appreciably for nearly a century (even with the growth of agronomic research and universities). It was limited by N, and the subsequent increase in yields mirrors our adoption of N fertilizers following World War II.[12]

Sanderman et al.[13] estimated that the total SOC released to atmospheric CO_2 over the duration of the history of farming is about 130 Gt C. Today, society releases about 10 Gt $C \, y^{-1}$ through combined industrial emissions and land use change,[14] which means that the historical loss of SOC amounts to more than a decade of present fossil fuel emissions to the atmosphere. As the Earth's greenhouse gas levels reach a point where societally disruptive

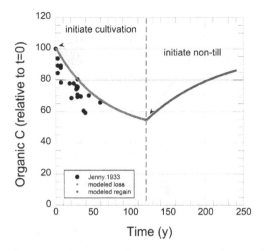

Figure 10.6 Observed relative changes in soil C in the surface soil horizons as a function of years of cultivation (data from Jenny, 1931) compared with model estimates (Eq. (10.4)) driven by the increase in k needed to reduce steady-state C to roughly 50 percent of its original condition. Also shown is the temporal response of the cultivated soil to a reversion to its original soil C decomposition rate under a new non-tillage management regime.

climate change becomes likely, the possibility of returning this 130 Gt of C to the soil becomes an appealing concept. To do so, from an inspection of Eq. (10.3), we can see that this requires some combination of decreasing k_f and/or increasing I_f. For example, if the soil examined earlier, which lost ~40 percent of its original C through cultivation (through an increased k_f), was placed into a no-till farming strategy with high inputs of residue, the C content of the soil might begin to increase (Figure 10.6). As with C loss, the rate of C return to the soil declines precipitously with time, and after a few decades it is a minor sink of C (Figure 10.6).

As global greenhouse gas emissions continue to alter the climate system, it is useful to at least consider the magnitude of potential C storage in soils through either deferred losses due to land use change or improvements in management that can restore some of the gigatons of C lost through past land use. These potential rates of sequestration are in some ways maximum values, because to achieve them requires economic and policy changes that will inspire their implementation. In addition, as discussed later in this chapter, these soil storage mechanisms will all be impacted by changing climates, which in turn cause losses of C via enhanced decomposition rates.

In a recent estimate, soil C mitigation (through either avoided emissions or enhanced sinks) could be up to 2.8 Gt CO_2 y^{-1} (1 Gt CO_2 = 0.27 Gt C).[15] The rates of soil vs. vegetation in different sectors (forest, agricultural, wetlands) is illustrated in Figure 10.7. The key challenge ahead will be the process of implementing the adoption of strategies to accomplish C sequestration that approaches these potential rates.

In principle, SOC can be returned to soils through changes in processes in the mass balance equations introduced earlier. However, the analysis here shows some of the potential pitfalls: (1) the rate of C return to soil declines rapidly with time (Figure 10.6) and (2) if management practices revert to other approaches, the loss of restored C is rapid,

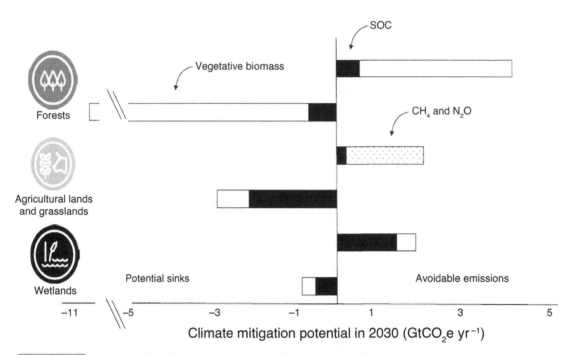

Figure 10.7 The bars to the left represent sinks, while those to the right represent avoided C losses. The dark part of the bars is the contribution of soil C to these activities, while the white part represents vegetation. From D. Bossio et al., The role of soil carbon in natural climate solutions, *Nature Sustainability*, doi:10.1038/s41893-020-0491z (2020).

which can cause sequestered C to be lost. In addition, as discussed later, increasing temperatures increase the decomposition rate k, and simply maintaining present amounts of C in soil will become increasingly difficult as the climate continues to respond to greenhouse gas increases in the atmosphere.[16]

10.2.2 The Effect of Irrigated Farming on Soil Biogeochemistry

The example we have just concluded focused on one soil property: SOC. As the importance of C on a global scale indicates, this is of course a serious issue. However, there are many other ways farming changes soil properties, and here the focus is on irrigated agriculture. The reason for this is that irrigation water is usually added to soils that initially formed under semiarid or arid conditions for thousands of years. The sudden change in hydrology and biology imposed by the addition of water is sometimes enormous. Table 10.1 gives an example of the environmental boundary conditions (e.g. state factors) for soil processes for an undisturbed and irrigated soil in the Great Central Valley of California. To examine these impacts, and to illustrate how the data sources and methods that have been covered in the book can be used, a comparison between a native and an irrigated soil in California is utilized. This comparison, conducted in 1992 by Natural Resource Conservation Service (NRCS) scientists,[17] is one of a limited number of such comparisons in the NRCS database.

Table 10.1 Approximate initial and post-farming environmental and biogeochemical boundary conditions for soils in the southern San Joaquin Valley of California

Property	Natural	Irrigated
MAP/irrigation (mm)	310	>1000
Average soil temperature @ 50 cm	16	~11–12
Biota	Annual grasses, forbs	Fruit and nut orchards, vineyards, alfalfa
Net primary production (kg ha^{-1} y^{-1})	1000–2000	15,000 (alfalfa)
N addition (kg ha^{-1} y^{-1})	10	200–300
S addition (kg ha^{-1} y^{-1})	<3	3000–30,000 (frequency of additions variable)

The original soil (Jerryslu series: an Aridisol with opal-cemented duripan and a sodium and clay-rich B horizon) is typical of a suite of soils that forms at the boundary between gently sloping alluvial fans from the Sierra Nevada (largely granitic alluvium) and the adjacent floodplains of rivers and drainage systems that slowly migrate northward toward the San Francisco Bay. Over the Holocene epoch (~10 Ky), a natural shallow groundwater table (comprised of relatively dilute waters equilibrated with granitic mineral constituents) has evaporated, leaving high concentrations of salt in the soil and raising the pH to 9 or higher.

The chemistry of a representative well from the region (~100 m deep) is provided in Table 10.2. The water is low in total solutes (low electrical conductivity) and is dominated by Na^+ and HCO_3^-, as is expected for water that has reacted with granitic alluvium from the nearby Sierra Nevada (quartz, Na/Ca plagioclase, biotite, hornblende). Evaporation over geological time has concentrated the Na bicarbonate in the soils (along with other solutes), raising the pH (Table 10.3). The high pH values have accentuated silicate mineral dissolution, helping to rapidly form clay minerals (including the unique zeolite mineral anaclime). The landscape is a mosaic of barren high-pH salt depressions (which form ponds in the winter) and slightly more elevated islands that support salt-tolerant grasses and shrubs. Due to the unavailability of better-adapted soils in the area for further agricultural expansion, these low-productivity soils have become slowly converted to agriculture, though the transformation to productive land takes some years, requiring time for irrigation water to remove salt and decrease the Na content (which reduces pH by removing $NaHCO_3$ and improves infiltration rates). The data we examine here were collected in 1992. As testament to the changing land use the planet is undergoing, the reference soil (Jerryslu) was itself converted to agriculture between 2017 and 2018. Thus, this natural reference site no longer exists.

Here, we explore changes in some selected soil properties: C and N, changes in physical properties (clay), and water-soluble salt composition. Concentrations can be converted to mass per unit land area (kg m^{-2}).

Table 10.2 Example of shallow groundwater chemistry in the east side of the San Joaquin Valley in Kern County, CA. USGS # 353332119223301

Conductivity	(dS/m)	0.274
CO_2	mmol/L	0.21
HCO_3		0.95
CO_3		0
NO_3		0.02
Ca		0.50
Na		1.48
Mg		0.01
K		0.02
Cl		0.89
SO_4		0.31
SiO_2		0.35
Sum +		2.52
Sum −		2.47
pH		7

First, the distinctive depth profile of clay in the Jerryslu (low at the surface with a bulge at 20 cm) has been largely homogenized by tillage and deep ripping of the soils (Figure 10.8a). This physical disruption and mixing improves soil water infiltration rates. Second, due to the high mean annual temperatures, combined with the present low biological productivity, the undisturbed soil has low C and N contents (Figure 10.8b, c). As Table 10.1 indicates, farming increases annual water inputs and likely reduces soil temperatures, which probably increases C inputs and reduces decomposition rates, resulting in higher C and N in the irrigated soils (Figure 10.6b, c).

Probably the most striking change in the soils is the change in the abundance and proportion of water-soluble ions. By (1) calculating the volume of each soil horizon, (2) multiplying by bulk density, and (3) multiplying by the mass of water per mass of soil at saturation, the total amount of water-soluble salt in each horizon (per m^{-2}) can be determined. By multiplying the reported concentration of each ion by it molar weight (and summing), the total mass of salt in the profiles of the native vs. irrigated soils can be compared. The upper 200 cm of the Jerryslu soil has 0.102 kg of salt per m^{-2} (or 1017 kg per ha). In contrast, the irrigated Atesh soil has 0.007 kg salt m^{-2}, or only 70 kg ha^{-1}. Thus, over the duration of irrigation (not known, but likely <~25 years), roughly 940 kg of salt per ha has been leached out of the upper 2 m of the soils and into the underlying vadose zone. Additionally, the ratios of elements have changed. In the native soil, Na/Ca ratios are very high (up to 270), and Cl>SO_4>HCO_3 in terms of anions. In contrast, in the irrigated soils the Na/Ca ratios are <4.9 and similar to the local irrigation waters (~3.0). In addition, HCO_3>SO_4>Cl in terms of the dominant anions (again reflecting irrigation waters). The loss of Na salts (bicarbonate) has also lowered the soil pH from >10 to 8 or less in the irrigated soils.

Table 10.3 The water-soluble (saturation extract) data, the water content at saturation, and the bulk density data for the Jerryslu and Atesh soils

Jerryslu #	Depth	Horizon	Ca	Mg	K	Na	CO$_3$	HCO$_3$	F	Cl	SO$_4$	NO$_3$	H$_2$O	BD
	cm		mmol L^{-1}											g cm^{-3}
93P01954	0–3	A	11.1	2.1	2	17.5		16.9	2.6	7.2	2.8		48.8	1.97
93P01955	3–8	E	5.3	0.9	1	29.7		9	0.5	19.9	5.9		28.1	1.67
93P01956	8–10	E2	2.2	0.3	1.6	190.6		10.4		113.3	54.9		26.6	1.8
93P01957	10–23	Btkn1	0.8	0.2	1.1	216.8	24.5	11.3		127.2	49.2		38.1	1.76
93P01958	23–31	Btkn2	1	0.3	0.4	122.8	13.6	9.1		68.1	28.6		38.6	1.51
93P01959	31–42	Btkn3	0.6	0.2	0.2	128.6	13	9.1		72.7	29.7		51.3	1.3
93P01960	42–74	Btkn4	0.6	0.2	0.2	162	16.8	11.3		87.4	40.5		54.9	1.33
93P01961	74–89	Btkn5	0.9	0.3	0.3	156.6	11.6	11.1		80.6	37.2		40.6	1.54
93P01962	89–107	Btkqm1	1.3	0.4	0.2	87		3.7		63.1	20.5		30.6	1.57
93P01963	107–126	Btkqm2	9.7	2.4	0.1	89		1.1		76.6	21.5		28.7	1.78
93P01964	126–140	Btkqm3	9.6	2.4	0.1	48		1		51.8	7.7		37.1	2
93P01965	140–163	Btk	9.5	2.2	0.1	26.6		1.1		35.7	2.6		40	1.86
93P01966	163–183	2C1	7.1	1.6	0.1	10.9		1		16.7	1.2		66.8	1.84
93P01967	183–200	2C2	4.3	1	0.1	4.5		0.9	0.1	9.2	0.3		27.9	1.78

Atesh	Depth	Horizon	Ca	Mg	K	Na	HCO$_3$	F	Cl	SO$_4$	NO$_3$	H$_2$O	BD
	cm		mmol L^{-1}										g cm^{-3}
93P01968	0–5	Ap1	9.4	1.4	0.1	4.8	12	0.1	0.9	1.8	0.2	38.1	1.53
93P01969	5–15	Ap2	4.3	0.6	0.1	2.7	4.8	0.1	0.4	0.7	0.5	36.2	1.54
93P01970	15–36	C1											1.52
93P01971	36–58	C2	1.4	0.4		4.4	2.9	0.2	0.8	1.3	0.2	38.6	1.4
93P01972	58–71	Btkn1	1.1	0.3		5.3	3.2	0.4	0.7	1.5	0.1	41.2	1.39
93P01973	71–91	Btkn2	1	0.3		4.4	2.4	0.4	0.8	1.7		30.5	1.67
93P01974	91–109	Btkqmb1	1	0.3		4.1	2.3	0.5	0.7	1.4	0.1	31.4	1.83
93P01975	109–124	Btkqmb2	1	0.3		4.1	2.4	0.4	0.6	1.3		37.6	1.6
93P01976	124–147	Btkqb1	0.9	0.3		4.2	2.1	0.3	0.9	1.6		36.9	1.77
93P01977	147–180	Btkqb2	1.2	0.3		4.6	2	0.3	1.3	2.3		32.8	1.9
93P01978	180–211	Btkqb3	1.8	0.4		5	1.7	0.1	1.9	2.7	0.1	32.7	1.58
93P01979	211–234	Btkb	2.2	0.5	0.1	5.5	1.6	0.1	2	4	0.1	29.4	1.8

BD = bulk density.

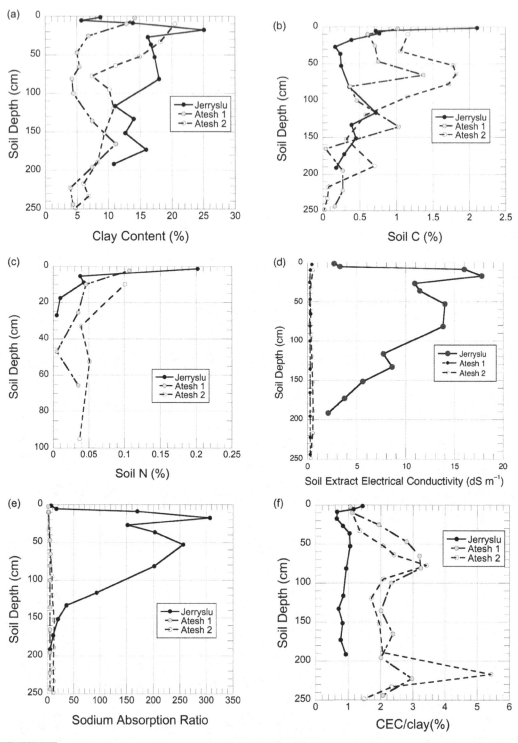

A comparison of soil properties in an undisturbed soil (Jerryslu) and two domesticated versions (Atesh) in irrigated farmland of Tulare County, CA. (a) Clay content, (b) soil C, (c) soil N, (d) electrical conductivity of saturation extracts, (e) sodium adsorption ratio of saturation extracts, and (f) CEC/clay ratios.

A somewhat unusual mineral in these soils is the zeolite mineral anaclime. Zeolites form in high pH conditions with abundantly available Si and cations (such as Na) and are thus expected in this unique environment. Zeolites, in general, are known for their high cation exchange capacity (CEC) (and are used in industrial settings due to their adsorption capacities) – up to 400 meq/100 g mineral. However, there are challenges to measuring the CEC of zeolites due to displacement of adsorbed cations – particularly the ability of NH_4 to displace Na in some species.[18] The somewhat surprising doubling of CEC/g of clay in the irrigated soils may be due to the removal of Na from the minerals and its replacement by Ca or other cations. This may subsequently cause the minerals to have a higher apparent CEC using the standard NH_4 acetate method employed for soils. This is simply a hypothesis. But it illustrates the discovery of unanticipated phenomena through simple data analysis and the generation of hypotheses for follow-up biogeochemical research.

There are relatively limited sets of paired soil studies (undisturbed vs. domesticated) in many agricultural landscapes. In the USA Soil Taxonomy, Arents are Entisols caused by human activity. A search of the Soil Characterization database reveals that more than 90 have been examined. Some of these are constructed soils (such as mine spoil rehabilitation) that have no undisturbed counterparts. However, there are maybe a few dozen others that have recognized natural starting points, so that investigations of human impacts on soils can be potentially conducted elsewhere in the USA.

10.2.3 Soil Carbon and Climate Change

It is now apparent to scientists that humans have increased atmospheric CO_2 through the combined effects of agriculture (see earlier) and fossil fuel emissions. Additionally, soil scientists recognize that changes in management (with key processes or levers described by the soil C balance equation, Eq. (10.3)) can lead to some removal of CO_2 from the atmosphere and its storage again in soil. Yet, Chapter 7 emphasizes that SOC is not a passive reservoir of organic C and is subject to changes due to corresponding changes in its surroundings, particularly temperature. In Chapter 7, using data from a recent paper, it was shown that soil C decomposition rates increase with mean annual temperature (MAT). In Figure 10.9, this relationship is illustrated. As introduced in Chapter 7, the inverse of decomposition rate is soil C residence time. The data in Figure 10.9, when converted to residence time, result in values nearly identical to that found in an independent global analysis of soil C storage and vegetation inputs to soil.[19] The derivative of the model fit in Figure 10.9 gives a temperature sensitivity of the decomposition constant k as $e^{(0.075223*MAT)}$.

These relationships can be viewed schematically in a cause–effect diagram (Figure 10.10). The result of one positive relation (increasing CO_2 increases mean annual temperature (MAT)) and two negative relationships (increases in MAT decrease SOC, and decreasing SOC increases atmospheric CO_2) result in a positive feedback loop, one by which increasing atmospheric CO_2 initiates a process that is self-sustaining. It should be noted that this diagram does not include potential changes in plant inputs to soils, which are uncertain. Thus, the focus is on the decomposition part of the soil C balance. How can one begin to

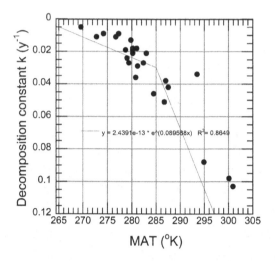

MAT (°K)

Figure 10.9 The correlation between the decomposition constant and mean annual temperature for sites in Table 7.1.

quantify the rates between the pools? Specifically, how does increasing temperature impact the global storage of soil organic matter?

The approach is to use a version of Eq. (10.1). The data we need are: (1) soil organic matter decomposition rates and response in the rate to temperature (which is given earlier), (2) soil C pools in different climate zones, and (3) projections of the rate of temperature change. As a simple estimate of soil C vs. climate zone, data from Post et al.[20] are used here and are provided in Table 10.4 (along with the value of k calculated from Figure 10.9). The global temperature has increased by 0.7 °C over the past 25 years,[21] a rate of 0.28 °C per year. Using these data and Eqs (10.4) and (10.5),[22] the yearly changes in the pool size of each soil climate zone due to slowly increasing temperature are calculated, and the results are summed to obtain yearly values as well as integrated effects over time. The subscript t refers to a specific year, and k_f (and if desired, I) is adjusted for each time step based on data and temperature changes. The result for a simple scenario, from 1990 to 2015, is illustrated in Figure 10.11. This interval is chosen to compare with an independent study of how soil respiration rates (and SOC decomposition) have changed over this same time interval. In the calculation, one assumption is that only half of soil C in the upper 1 m cycles on decadal timescales (thus reducing reactive C in each climate pool by 50 percent). This reduction in rapid cycling pool size is intended to recognize that not all soil C cycles on decadal timescales and thus provides a more conservative estimate of the net soil response to temperature change. The results suggest that after a 25-year period, during which temperatures increased in all ecosystems by 0.7 °C, global soils released roughly 0.56 Gt of C per year due to changes in decomposition rates. Over the duration of this period, soils apparently released 8.3 Gt of C. This calculation, intended to illustrate some simple principles, is essentially identical to an independent approach in a study suggesting that global heterotrophic soil respiration has increased by 1.2 percent over the past 25 years. If

Table 10.4 Soil C storage by world life zone

Life zone	Area (10^{12} m^{-2})	Mean soil C (kg m^{-2})	MAP (mm)	MAT (°C)
Tundra	8.8	21.8	500	2.25
Boreal desert	2	10.2	125	4.5
Cool desert	4.2	9.9	125	9
Warm desert	14	1.4	125	22
Tropical desert bush	1.2	2	188	25
Cool temperate steppe	9	13.3	375	9
Temperate thorn steppe	3.9	7.6	375	14.5
Tropical woodland and savanna	24	5.4	375	23.5
Boreal forest – moist	4.2	11.6	375	4.5
Boreal forest – wet	6.9	19.3	1250	4.5
Temperate forest – cool	3.4	12.7	2250	9
Temperate forest – warm	8.6	7.1	4250	14.5
Tropical forest – very dry	3.6	6.1	750	28
Tropical forest – dry	2.4	9.9	1250	23.5
Tropical forest – moist	5.3	11.4	2500	23.5
Tropical forest – wet	4.1	19.1	6000	23.5
Globe	129.6	10.8		

MAP = mean annual precipitation.

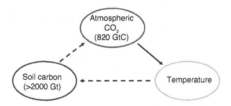

Figure 10.10   A schematic cause–effect relationship between soil C and the climate system. Dashed lines indicate a negative relationship while a solid line indicates a positive relationship.

we take the standard estimate of global soil respiration (about 60 Gt C y^{-1}), this amounts to approximately 0.7 Gt C y^{-1}, nearly identical to the calculated temperature sensitivity of soil C to increasing temperatures obtained here.

The tools developed here are embedded (in a variety of permeations) into complex climate models, along with large data sets used to examine questions regarding soil response to climate change and ways that soils can be managed to reduce additional emissions. While some approaches involve multiple pools of C, the underlying principles are similar to the introduction here. The approaches here are easily adaptable for a number of first-order, back-of-the-envelope-type investigations. However, these simple calculations are generally consistent with peer-reviewed research, which reveals that soil–climate feedbacks are likely already contributing to our global climate system.

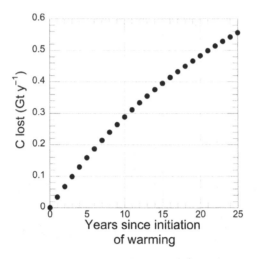

Figure 10.11 An estimate of the global soil C pool response to a 0.03 $C\,y^{-1}$ increase for 25 years. Each life zone pool in Table 10.4 was given an initial decomposition and input rates based on relations in Figure 10.9. Using Eq. (10.3), the changes in storage in each life zone were calculated and then summed.

10.2.4 Soil Erosion

An examination of Figure 10.2b suggests that the most significant, and nearly permanent, impact of farming on soils is erosion. In addition to removing the physical matrix itself, it removes organic matter and SOC, and greatly impacts the nutrient cycles. As the figure illustrates, soil erosion can drive many of the key soil cycles into a non-steady-state condition, one that is unsustainable without significant external inputs.

As discussed in the previous chapter, natural soil erosion on vegetation-mantled hill-slopes is a diffusive-like process known as creep. The rate is dependent on slope and a local rate constant that incorporates the local climate, vegetation, lithology, and animal life that help drive the movement of the soil:

$$q_s = -K\frac{dz}{dx} \tag{10.5}$$

Chapter 9 showed how the production rate of soil, and in some locations the erosion mechanism of soil, is linked directly to soil thickness, creating two negative feedback loops that sustain soil thickness, in natural undisturbed settings, in the face of perturbations.

Once the natural vegetative cover is removed by tillage and soil is exposed to particle movement by flowing water, the mechanism of soil movement changes from diffusive to advective. This is a profound change. Predicting and managing soil erosion under highly altered conditions has been a major effort for the past century, yet there remain significant challenges and opportunities to (1) developing new types of models that are available to quantify erosion and (2) quantifying how much erosion is allowable to maintain a sustainable soil system.

Briefly, the processes that cause advective soil erosion begin with raindrop impact, a subsequent dislodgment of soil particles, the removal of these particles in moving water as sheet and rill flow, and the subsequent deposition of mobilized material as hydrological conditions change with topography. In the USA, two very different approaches have been developed to predict and estimate erosion under altered conditions: (1) an empirical approach known as the Revised Universal Soil Loss Equation (RUSLE) and (2) the process-based Water Erosion Prediction Project (WEPP). The WEPP model, which is mechanistic, requires a large set of input parameters, which can be accessed and adjusted through the online versions of the model, allowing users to explore erosion rates on a local scale. In contrast, the RUSLE is defined as

$$A = R * K * LS * C * P \tag{10.6}$$

where A = estimated soil loss in tons per acre per year, R = rainfall erosivity factor, K = soil erodibility factor, L = slope length factor, S = slope steepness factor, C = crop management factor, and P = practice support factor.

Geomorphologists have developed an empirical model that embeds within it some of the processes and empirical relations of both these models:[23]

$$q_s = dA^m S^n \tag{10.7}$$

where A = drainage area, S = local slope, and d, m, and S_n are fitted parameters (into which climate, etc., is embedded).

In the past decade or so, concern about water-driven soil erosion has again been raised, and efforts to estimate the global impact, and any temporal changes in the rate on both regional and global scales, have increased. Some of the prominent recent assessments have relied on the RUSLE approach. In one recent effort on global soil erosion, Borelli et al.,[24] using a RUSLE model coupled with geographic information system (GIS) analyses of land use, calculated that the annual global cropland erosion is 17 Pg y^{-1}, which is 12.7 Mg ha^{-1} y^{-1}. Erosion data can be expressed in units of length (as in Chapter 9) or mass. Here, to convert the mass estimates to units of length, we assume a soil bulk density of 1.5 g cm^{-3} (the reader can use other ranges in bulk density in their own calculations). After conversions, this rate is equivalent to 847 m My^{-1}. This rate is much larger than the author's estimate of the overall global soil erosion rate (including forests, pastures, etc.) of 187 m My^{-1}, illustrating the acceleration of erosion due to farming practices.

The irony of the present research on soil erosion, and its impacts on sustainability, is that the concept and quantity of *allowable or sustainable* rates of soil erosion are based on anecdotal and poorly defined evidence. The "tolerable soil loss," or the term T in the RUSLE, is defined as the maximum rate of soil loss that still permits a high level of crop productivity to be sustained indefinitely. Borelli et al. use a value of 10 Mg ha y^{-1}, which is 667 m My. T values for soils in the USA can be found in soil survey reports and are commonly 11.2 Mg ha y^{-1}.[25]

Are these assumed tolerable ranges in soil erosion correct? The key is that erosion rates do not exceed rates of soil production. As mentioned in the previous chapter, the rates of soil production on soil-mantled hillslopes have only recently begun to be quantified on a

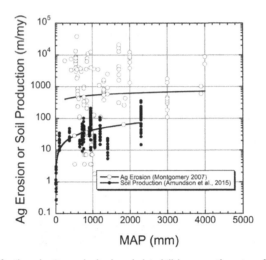

Figure 10.12   Natural rates of soil production on bedrock-underlain hillslopes vs. the rates of agricultural soil erosion as a function of mean annual precipitation. From R. Amundson et al., Hillslope soils and vegetation, *Geomorphology*, 234: 122–132 (2015).

large scale. It should be noted that many of these soil production studies have been conducted on hillslope soils overlying bedrock, and soil production on soft sediments – where much farming occurs – is less studied. The studies on bedrock reveal that natural rates of soil production in areas with rainfall suitable for crop production range between 10 and 200 m My^{-1}, rates sometimes orders of magnitude lower than observed agricultural erosion in the same climate zones (Figure 10.12), and factors of 1/70 to 2/7 of the "tolerable" soil erosion rates. The underlying importance of these discrepancies is that accelerated erosion can rapidly impact the thickness of the soil. The time frame for the removal of the entire soil thickness is[26]

$$T_c = \frac{H}{E - P} \tag{10.8}$$

where T_c = the time to erode through a soil of thickness H, E = erosion rate, and P = production rate.

The loss of thick, dark A horizons from many steeply sloping agricultural fields is due to the fact that erosion rates now are large relative to the biological rates of C and N inputs and cycling.

In a recent study in western Minnesota, Jelinski et al.[27] used ^{10}Be and ^{137}Cs to develop estimates of pre- and post-cultivation erosion rates on gentle paired uplands formed on glacial till. The work revealed that pre-cultivation erosion rates were about 50 m My, while the post-cultivation erosion rates (for the toposequence considered here) ranged from 290 to 4900 m My, orders of magnitude higher than natural background levels. This recent acceleration in erosion rates likely exceeds the pace of most soil processes. For example, Jelinski and Yoo[28] mapped the distribution of eroded phase soils in the USA (Figure 10.13).

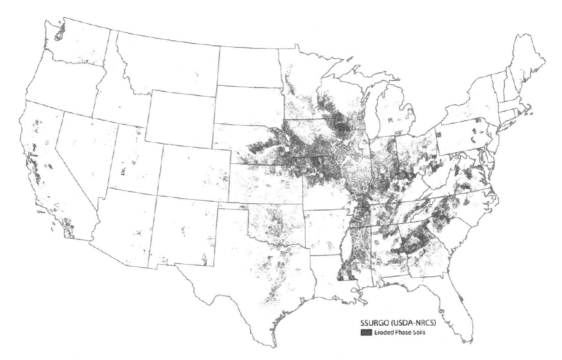

Figure 10.13 A map of eroded phase soils in the conterminous USA. In the eroded phase soils, soil morphologic changes due to accelerated soil erosion are evident. The reference for each eroded phase soil is an identifiable layer or feature. Figure from N. A. Jelinski and K. Yoo, The distribution and genesis of eroded phase soils in the conterminous United States, *Geoderma*, 279: 149–164 (2016).

Among other considerations, the authors noted that many areas now mapped as containing Inceptisols (soils without organic-rich A horizons) were likely Mollisols (dark humus-rich A horizons) prior to cultivation, and that erosion has proceeded faster than the rate at which these horizons could be restored by C cycling.

The Minnesota study provides a quantitative perspective of the changed C cycle. Noncultivated soils have between 16 and 21 kg C m^{-2} (whole profile depth), while cultivated soils range between 6.8 and 13.2 kg C m^{-2}, a reduction of 50 percent – which is consistent with the work of Jenny discussed in Chapter 7 (Table 10.5). If it is assumed that both cultivated and noncultivated sites are near steady state, it is possible to examine changed rates of C cycling (Table 10.5). Some very approximate C input rates are about 500 g C m^{-2} y^{-2} for row crops,[29] while for natural grassland, at the rainfall of western Minnesota, the net C inputs are about 200 g C m^{-2} y^{-1}.[30] Slightly differing rates will, of course, result in better site-specific information; however, these preliminary rates provide a first-order understanding of how agriculture has impacted the C cycle.

In Chapter 7, it was assumed that most C is lost from soil via respiration, but in rapidly eroding soils this is incorrect, and net loss must include erosion. This erosive loss of soil C on cultivated fields is quickly transported to depositional settings where the C along with

Table 10.5 Summary of soil C, calculated erosion rates, and C cycling rate parameters for a cultivated and uncultivated soil toposequence in western Minnesota

Site	Total soil C ($kg\,m^{-2}$)	Erosion rate ($m\,My^{-1}$)	Slope	Soil C erosion rate ($g\,C\,m^{-2}\,y^{-1}$)	NPP ($g\,C\,m^{-2}\,y^{-1}$)	Soil C residence time (y)	k_e (y^{-1})	k_r (y^{-1})
Cultivated								
	6.8	3600	0.1	49.7	500	13.6	0.007	0.07
	10.2	4920	0.11	72.0	500	20.3	0.007	0.04
	13.2	290	0.07	6.0	500	26.5	0.0005	0.04
Native								
	19.6	47	0.18	2.4	200	97.8	0.0001	0.01
	19.1	47	0.21	2.0	200	95.6	0.0001	0.01
	16.1	47	0.19	1.9	200	80.6	0.0001	0.01
	21.0	47	0.18	1.8	200	104.2	0.00009	0.01

Original data from N. A. Jelinski et al., Meteoric beryllium-10 as a tracer of erosion due to post-settlement land use in west-central Minnesota, USA, *JGR Earth Surface*, doi:10.1029/2018JF004720.
NPP = net primary production.

the soil material accumulates (Figure 10.14). The expression that describes this relationship is[31]

$$\frac{dC}{dt} = I - (k_r + k_e)C \tag{10.9}$$

where k_r and k_e are the decomposition constants for respiration and erosion, respectively.

The annual loss of soil C by erosion divided by the total C pool gives k_e, which is roughly an order of magnitude smaller than k_r (Table 10.5). In the cultivated landscapes, the soil C erosion rate constant is about a factor of 5 greater than that on the native landscape. Assuming steady state, the respiration rate constant is equal to the (NPP − erosion C loss)/total C. For the cultivated sites, the respiration constants are about an order of magnitude higher than for the native sites. This is due mathematically to the lower total C and the higher estimated rates of C inputs. In a practical sense, this is also reasonable in that cultivation physically disrupts soil and makes C more accessible to decomposing organisms.[32]

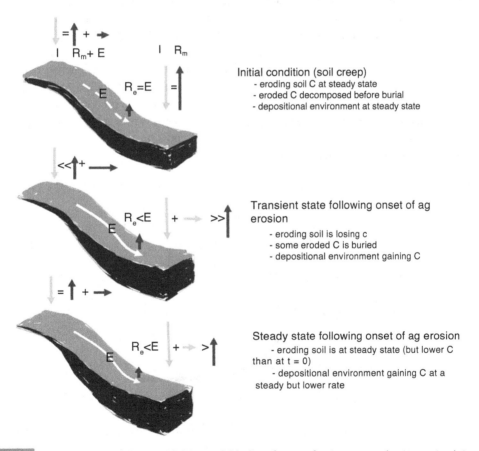

Initial condition (soil creep)
- eroding soil C at steady state
- eroded C decomposed before burial
- depositional environment at steady state

Transient state following onset of ag erosion
- eroding soil is losing c
- some eroded C is buried
- depositional environment gaining C

Steady state following onset of ag erosion
- eroding soil is at steady state (but lower C than at t = 0)
- depositional environment gaining C at a steady but lower rate

Figure 10.14 A representation of changes in hillslope soil C budgets from pre-farming to a steady-state erosional situation.

If we assume that the agricultural soils have lost 50 percent of their original C (based on data provided), then it is possible to explore the dynamics of the soil C in the years following the introduction of cultivation. The model to describe this is the solution to Eq. (10.7):[33]

$$C_s(t) = C_o e^{-(k_r+k_e)t} + \frac{I}{k_r + k_e}\left(1 - e^{-(k_r+k_e)t}\right) \tag{10.10}$$

where $C_s(t)$ and C_0 are the soil C at time t and t = 0, respectively. By using the I and k values for the first cultivated soil, the results (Figure 10.15) largely follow those of Figure 10.6, which were direct observations as opposed to the modeled results here.

The accelerated erosion of soil under agriculture removes SOC along with the mineral matter that is being transported. This perturbs the approximately steady-state soil C pool and transports C outside the soil boundary. What happens to the eroded C is possibly important to the global C budget. Stallard[34] suggested that some or much of the eroded soil C is rapidly transported to depositional settings where the decomposition rates are lower, or much lower, than they were in the original soil (Figure 10.14). Meanwhile, at the eroding soil sites, inputs (possibly increased through fertilization and management) replace some or all of the eroded C. Stallard calculated that the size of this annual erosive flux of C could be between 0.6 and 1.5 Gt y^{-1}. A more recent estimate places this flux at 0.5 Gt C y^{-1}.[35] If this C does not decompose and is instead buried and stored in sediments, it represents a significant unrecognized sink of soil and terrestrial C, and this has been proposed to be part of the global land sink of C that is presently removing a portion of anthropogenic CO_2 each year. On the other hand, if it largely decomposes over a short time span, it simply represents an acceleration of the soil C cycle and possibly a net anthropogenically mediated source to the atmosphere.

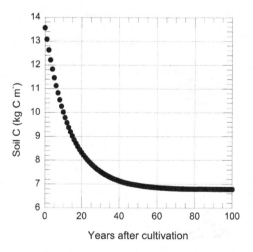

Figure 10.15 Modeled soil C loss following the onset of cultivation for a soil in western Minnesota using data in Table 10.5 and Eq. (10.10).

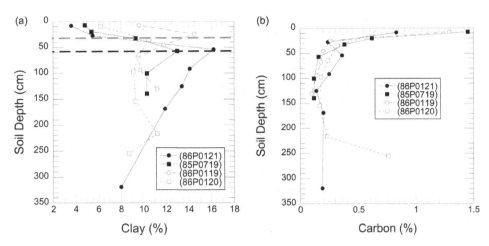

Figure 10.16 A comparison of two less eroded (P0121, P0719) vs. two more excessively eroded (P0119, P0120) Memphis soils from Tennessee. (a) Clay content and (b) organic C percent vs. soil depth. Data from NRCS database.

There are many unresolved and uncertain issues about soil erosion and soil sustainability, but new research allows soil scientists to retest old assumptions with new information and concepts. To do so, let's return to the Memphis soil from Chapter 5 (Figure 10.16). In the state of Tennessee, four profiles of the Memphis have been analyzed. Two appear to be largely unaffected by erosion and retain the A and E horizon sequence at the surface. The other two (in farming) no longer have an E horizon and have a thinner A horizon. If we compare the depth at which the maximum clay begins, two appear to have lost approximately 30 cm to erosion, and their Bt horizons are now close to the surface (Figure 10.16a). Quinton et al.[36] have estimated that erosion globally removes 17.5 Tg P y^{-1}, while 18 Tg P y^{-1} is added as fertilizer. If one examines undisturbed vs. cultivated/eroded soils in other locations, it is likely that due to fertilization, the P content of agricultural soils may indeed match or exceed those of their undisturbed counterparts. However, the annual loss due to erosion, which equals the global fertilizer inputs, represents an enormous flux of a nonrenewable resource.

Remarkably, despite evidence that two of the profiles have lost about 30 cm of soil due to erosion, the C content of the soils is remarkably similar (Figure 10.16b). As discussed earlier, this seems to conform to the hypothesis that C inputs and cycling on eroding landscapes can maintain a steady-state C stock (not always the same amount as the undisturbed state) while continuously funneling some C with eroding sediment to depositional settings, where it may be retained as an atmospheric C sink.

To conclude this introduction to soil erosion, more questions might arise (for both student and researcher) than answers. Some areas of still remaining uncertainty include the following:

• We require maps and better data on the legacy of early agricultural erosion in many parts of the world. How do the present soils in these regions compare with the preagricultural landscape, and how have the soils recovered from early human use?

- We need to link soil erosion data to the depth of the erodible material that mantles the landscape. What is the depth to bedrock or less desirable sediment?
- In terms of erosion on soil C, there is a lack of research on the fate of C once it is deposited in depositional settings. Almost certainly it decomposes more slowly, but at what rates? Without better constraints on these rates, we still will not be fully certain of the sign of erosional loss of C on the soil C budget.
- As with C, we need to know the fate of N, and since N is now commonly derived from fertilizer, it has impacts on aquatic systems and global atmospheric chemistry.
- It is clear that erosion removes P. However, erosion removes the surficial weathered soil mantle, which has lost some (in young and temperate settings) or most (in old and warm/humid settings) of its parent material P. What are the possible diverse spatial differences or impacts of erosion on P availability?

10.3 Summary

The soils of 25 percent or more of the Earth's surface bear the strong imprint of a sixth Factor of Soil Formation: us.[37] Human use of soils has not only changed their processes and properties; it has impacted our global biogeochemical systems. Science not only provides a way to understand our impacts on soils but also provides ways to change our activities so that we can create environmental conditions that ensure the long-term existence of soil resources and the processes they provide for our future.

This final chapter likely represents the best reason for studying and understanding soils: our own survival and our prosperity. Our planet has evolved for 4.5 billion years, resulting in multiple soil processes that are part of a complex set of interactions and feedbacks that maintain the global biosphere. We invented agriculture 10,000 years ago and imposed it on the Earth's landscape, fundamentally altering long-standing and complex feedbacks. It is a challenge to find ways in which we can carry on this practice indefinitely. Additionally, managing soils and their processes is part of a significant effort to potentially mitigate, and also to adapt to, global change.[38] Understanding soil biogeochemistry provides some of the intellectual tools needed to attend to these and other issues of importance to all of us.

10.4 Activities

10.4.1 Back of the Envelope Biogeochemistry Calculations

Chapter 7 dwelt on soil C loss by farming. If we assume a C:N ratio of 15:1 for cultivated soils, how many kilograms of N have been mobilized (lost as gas or in aqueous form) due to the global history of agriculture?

10.4.2 Using Web-Based Data

N fertilizer production is credited as being the first "Green Revolution" and is thought to be the process that allows us to support roughly half the world's population. Conduct a Web search on estimates of global N fertilizer use (make sure they are in, or you convert them to, units of N). How do our annual rates compare with the total deficit caused by farming?

10.4.3 Simple Climate Impact Calculations

For the soil considered in the questions in Chapter 7, assume (as indeed is likely) that it began warming three decades ago at a rate of $0.3\ °C\ y^{-1}$. Assuming that the decomposition constant is related to temperature by

$$k = e0^{.075223*MAT}$$

calculate the total C in the soil over the past 30 years, assuming it was at steady state initially.

10.4.4 Erosion Effects

The global rate of soil erosion is reported to be about $2.4\ t\ ha^{-1}\ y^{-1}$.

a. From data we have used(or from a web search), what is an approximate total P content of soil?
b. How many kilograms of P are removed each year by soil erosion?
c. How does erosional loss of P compare with annual rates of P fertilization (again, make sure fertilizer is in units of P)?

10.4.5 Land Use

There has long been concern about soil conversion to urban lands and our loss of its use for agricultural or other purposes. The use of soil for cemeteries is another use or disruption of soils, though, of course, funerary practices vary enormously by culture (and, probably, with recognition of the scarcity or value of land). If burial of people were the only funeral practice, roughly what would be the land area used annually (in square kilometers)? Compare this with the size of a state or nation of interest.

10.4.6 Sustainability of Soils

The global soil erosion rate is given earlier. Assuming that the average agricultural soil is about 1 m thick, how many years would it take for the average global soil to be removed by erosion if there were no compensating soil production? (Hint: you must assume a bulk density for eroded soil.)

Afterword

I began writing this book several decades ago. Several significant and unexpected life hurdles emerged along the way that delayed its completion. But, over time, I "salvaged" several early chapters for review articles: Amundson, 2001;[1] Amundson, 2003;[2] Amundson, 2014.[3] Portions of those articles are embedded here, along with many new changes in the field since those publication dates.

The best thing that happened during this long gestation period was the outstanding work of my graduate students, which enriches this book and has transformed my understanding of soils. I mention those whose work is directly linked to this book. Gene Kelly helped expose me to the soils of the Great Plains and introduced me to biomineralization. Stephanie Ewing, Justine Owen, Kari Finstad, and Marco Pfeiffer have collectively provided a quantum leap forward in our understanding of the driest desert on Earth, in northern Chile. As a group, they have contributed to understanding the climate threshold that exists between the biotic and abiotic parts of our planet – and its profound impact on soil and landscape biogeochemistry. Erik Oerter advanced the ability of soils to provide paleoclimate information from carbonate using micro-sampling and dating techniques. Troy Baisden introduced novel and original insights into soil C and N isotopes and cycling. Jon Sanderman's work on soil C cycling research continues to inform and advance this important field. My interaction with Jennifer Mills has inspired a wealth of ideas, and she also read and commented on a preliminary version of the book. Finally, Kyungsoo Yoo's research on hillslope soil processes was the first to link soil biogeochemistry with the physical processes of soil production and transport, and his work continues to illuminate these processes. The book is shaped by the wealth of their ideas.

I owe also an enormous debt to colleagues. Lanny Lund took a chance on me as a graduate student and helped me figure out how to set up my first independent research project. The late Gordon Huntington and Oliver Chadwick honed my field science skills. Oliver also illuminated my understanding of desert soils, the part of the Earth that still mainly attracts me. Thanks to both of you. I have learned about entirely new fields of soil biogeochemistry and Earth surface processes from Bill Dietrich, Arjun Heimsath, and Kuni Nishiizumi. Jill Banfield and her lab group, especially Jacob West-Roberts, who reviewed the biology chapter, have been gracious enough to help my students and me begin the process of bridging soil geochemical research with the frontiers of microbiology. Mic DeNiro was the one who provided support for my ideas about the use of isotopes in soils and plants before most others and went along for an interesting and productive ride. He also taught me the laboratory side of isotope biogeochemistry, a skill as hard earned and as valuable as the experience of months or years in the field. Thank you.

I was fortunate, in my first six years at Berkeley, to be able to meet frequently with Hans Jenny. In his mid-eighties at the time, he would summon me to his office for Socratic interrogations of my view of various topics relevant to soils and nature. It wasn't always relaxing, but it was interesting, and now, decades later, I am much the better for it. When I, or many people across the globe, design field experiments, his vision guides our questions and how we go about answering them. There is no other scientist in the field of soil biogeochemistry whom young scientists should study as a starting point for their research.

But, in the end, I wouldn't be writing this book but for the Fates that decided to cast me as the son of a farm family in South Dakota. When I was growing up, farming was a back-breaking and (potentially) spirit-bending occupation. It likely still is. I had a Vocational Agriculture teacher, Mr. William Bryant, who captured my attention regarding the critical importance of soil conservation when I was just a high school freshman. Without him, I never would be doing what I am. I thank my parents for, eventually, accepting my life-altering choice to leave South Dakota and to follow a different path than they had. This choice changed my life, much for the better I think. This is likely the most important lesson I have learned: let your children or those you mentor choose what they wish to do with their own lives. No one knows where those paths might lead.

Notes

1 Introduction to Soils

1. C. Darwin, *The Origin of Species by Means of Natural Selection or the Preservation of Favoured Races in the Struggle for Life*, Penguin Books, London, England (1985), p. 107.

2. From Greek *pédon* ground, -logy from Greek *–logia* discourse, *The Oxford Dictionary of English Etymology*, C. T. Onions, G. W. S. Friedrichsen, and R. W. Bunchfield (eds), Oxford University Press, London (1966).

3. The origin of the word "pedology" has been attributed to the German scientist F. A. Fallou in his 1862 book *Pedologie oder allgemeine und besondere Bodenkunde* by J. P. Tandarich and S. W. Sprecher, The intellectual background for the factors of soil formation, pp. 1–13, in: R. Amundson et al. (eds), *Factors of Soil Formation: A Fiftieth Anniversary Retrospective*, Soil Science Society of America, Madison, WI (1994); see also *Oxford Dictionary*.

4. For discussion of the development of pedological concepts, see H. Jenny, *E. W. Hilgard and the Birth of Modern Soil Science*, Industrie Grafiche V. Lischi & Figli, Pisa (1961); Tandarich and Sprecher, The intellectual background; D. H. Yaalon and S. Berkowicz (eds), *History of Soil Science. International Perspectives*, Catena Verlag GMBH, Reiskirchen, Germany (1997).

5. H. Jenny, *Factors of Soil Formation. A System of Quantitative Pedology*, Dover Publications, New York (1994), chapter 1.

6. Jenny, *Factors of Soil Formation*.

7. Darwin, *Origin of Species*.

8. L. P. Villarreal, Are viruses alive? *Scientific American*, 292: 100–105 (2004).

9. Jenny, *Factors of Soil Formation*; H. Jenny, Derivation of state factor equations of soils and ecosystems, *Proceedings – Soil Science Society of America*, 25: 385–388 (1961).

10. V. V. Dokuchaev, *Russian Chernozem: Selected Works of V. V. Dokuchaev. Vol. 1*, translated by the Israel Program for Scientific Translations, Jerusalem (1967); see discussion by R. Amundson and H. Jenny, On a state factor model of ecosystems, *Bioscience*, 47: 536–543 (1997).

11. R. Amundson, Foreword to *Factors of Soil Formation,* in: H. Jenny (ed), *Factors of Soil Formation*.

12. Jenny, *Factors of Soil Formation*.

13. G. Sposito, *Chemical Equilibria and Kinetics in Soils*, Oxford Press, New York (1994).

14.   A. G. Tansley, The use and abuse of vegetational concepts and terms, *Ecology*, 16: 284–307 (1935).

15.   Jenny, *Factors of Soil Formation*.

16.   This is one of the most misunderstood aspects of the model, with the common misunderstanding being that factors are independent of *each other,* which may or may not be true for a given location.

17.   P. W. Birkeland, *Soils and Geomorphology*, 3rd Edition, Oxford University Press, New York (1999).

18.   R. Amundson, J. Harden, and M. J. Singer (eds), *Factors of Soil Formation. A Fiftieth Anniversary Retrospective*, SSSA Special Publication No. 33, Soil Science Society of America, Madison, WI (1994).

19.   R. Amundson and H. Jenny, On a state factor model of ecosystems, *Bioscience*, 47: 536–543 (1997).

20.   C. D. Koven et al., Higher climatological temperature sensitivity of soil carbon in cold than warm climates, *Nature Climate Change*, doi: 10.1038/NCLIMATE3421 (2017).

21.   H. Jenny, Relation of climatic factors to the amount of nitrogen in soils, *Journal of the American Society of Agronomy*, 20: 900–912 (1928).

22.   Highway 61, and the natural gradient it follows, is central to American music of the 20th century. The blues legend Robert Johnson is reported, by one account, to have sold his soul to the Devil in order to play the guitar at the intersection of Highways 61 and 49 in Clarksdale, MS. Much later, Bob Dylan released the revolutionary album *Highway 61 Revisited*. Graceland, and its soil, which is discussed in Chapter 5, is a few miles east of Highway 61 in Memphis. From a biogeochemical perspective, the nature of the loess along the eastern Mississippi has been a long-time focus of research and has received recent and notable renewed biogeochemical analysis, with papers by D. R. Muhs et al., Impact of climate and parent material on chemical weathering in loess-derived soils of the Mississippi River Valley, *Soil Science Society of America Journal*, 65: 1761–1777 (2001) and J. Z. Williams et al., The temperature dependence of feldspar dissolution determined using a coupled weathering-climate model for Holocene-aged loess soils, *Geoderma*, 156: 11–19 (2011). Soil erosion and C cycling on the loess soils in Mississippi were investigated in an important paper by J. W. Harden et al., Dynamic replacement and loss of soil carbon on eroding cropland, *Global Biogeochemical Cycles*, 13: 885–901 (1999).

23.   I. C. Burke et al., Texture, climate, and cultivation effects on soil organic matter content in U.S. grassland soils, *Soil Science Society of America Journal*, 53, 800–805 (1989); E. F. Kelly et al., Stable carbon isotope composition of carbonate in Holocene grassland soils, *Soil Science Society of America Journal*, 55, 1651–1658 (1991).

24.   H. Jenny, A study of the influence of climate upon the nitrogen and organic matter content of the soil, Missouri Agricultural Experiment Station Bulletin, 152 (1930).

25.   Muhs et al., Impact of climate and parent material.[23]

26.   H. Jenny, An equation of state for soil nitrogen, *Journal of Physical Chemistry*, 34: 1053–1057 (1930).

27.   H. Jenny and C. D. Leonard, Functional relationship between soil properties and rainfall, *Soil Science*, 38: 363–381 (1934).

2 An Overview of the Biogeochemistry of Soils

1. T. Ferris, Seeing in the dark, *The New Yorker*, August 10, pp. 55–61 (1998).

2. Chapter 7 in C. Allègre, *From Stone to Star. A View of Modern Geology*, Harvard University Press, Cambridge (1992); also chapter 1 in N. N. Greenwood and A. Earnshaw, *Chemistry of the Elements*, 2nd Edition, Butterworth-Heinemann, Oxford (1997). See also: W. H. Schlesinger and E. Bernhardt, *Biogeochemistry. An Analysis of Global Change*, 4th Edition, Academic Press, Waltham, MA (2020).

3. Chapter 1 in Greenwood and Earnshaw, *Chemistry of the Elements*.

4. A discussion of the difficulties facing a formulation of "average" crust are presented by S. R. Taylor and S. M. McLennan, *The Continental Crust: Its Composition and Evolution*, Blackwell Scientific, Oxford (1985). Here I have used the estimated average composition of the upper crust (upper 10 km) from Taylor and McLennan, supplemented with data on additional elements from table 3.3 in H. J. M. Bowen, *Environmental Chemistry of the Elements*, Academic Press, London (1979).

5. G. H. Brimhall, Preliminary fractionation patterns of ore metals through earth history, *Chemical Geology*, 64: 1–16 (1987).

6. Chapter 8 in Allègre, *From Stone to Star*.

7. The mineral illustrations presented here are from K. Momma and F. Izumi, VESTA 3 for three-dimensional visualization of crystal, volumetric, and morphology data, *Journal of Applied Crystallography*, 44: 1272–1276 (2011). Another resource is P. Barak and E. A. Nater, 1997–2018, The Virtual Museum of Minerals and Molecules, online resource, http://virtual-museum.soils.wisc.edu, and a key source of more detailed information on the mineralogical composition of soils, particularly the secondary minerals, is J. B. Dixon and S. B. Weed (eds), *Minerals in Soil Environments*, 2nd Edition, Soil Science Society of America, Madison, WI (1989).

8. An example of a recent paper exploring this is H. W. Scherer et al., Ammonium fixation and release by clay minerals as influenced by potassium, *Plant Soil Environment*, 60: 325–331 (2014).

9. The recognition that rainfall played a strong role on the depth to the beginning of a carbonate layer was observed by H. Jenny and C. D. Leonard, Functional relationships between soil properties and rainfall, *Soil Science*, 38: 363–381 (1934), who drove across the Great Plains observing how a number of soil properties responded to rainfall. A download of soil data from the NRCS archive by D. Royer, Depth to pedogenic carbonate horizon as a paleoprecipitation indicator? *Geology*, 27: 1123–1126 (1999) showed little if any relationship, likely due to an array of other soil-forming factors. G. J. Retallack, Pedogenic carbonate proxies for amount and seasonality of precipitation in paleosols, *Geology*, 33: 333–336 (2004), in a more selective data collection and analysis, showed a number of carbonate relations to climate. The history and controversy illustrate that the impact of one state factor on a soil property can best be observed through judicious experimental design and site selection in order to reduce compounding variables.

10. E. W. Selssarev et al., Landscape age as a major control on the geography of soil weathering, *Global Biogeochemical Cycles*, doi: 10.1029/2019GB006266.

11. For a recent, and much more comprehensive, coverage on many aspects of mineral weathering please refer to volume 7 (Surface and Groundwater, Weathering and Soils) in: H. D. Holland and K. K. Turekian, *Treatise on Geochemistry*, 2nd Edition, Elsevier, Oxford (2014).

12. P. J. Frings et al., The continental Si cycle and its impact on the ocean Si isotope budget, *Chemical Geology*, 425: 12–36 (2016).

13. It is worth repeating that these are coarse comparisons with still uncertain values for many data sets.

3 The Biology in Soil Biogeochemistry

1. V. Vernadsky, *The Biosphere*, Copernicus, Springer-Verlag, New York (1926).

2. L. A. Hug et al., A new view of the tree of life, *Nature Microbiology*, 1: 16048 (2016).

3. L. Sagan, On the origin of mitosing cells, *Journal of Theoretical Biology*, 14: 225–274 (1967).

4. R. G. Joergensen, J. Wu, and P. C. Brookes, Measuring soil microbial biomass using an automated procedure, *Soil Biology and Biogeochemistry*, 43: 873–876 (2011).

5. X. Xu et al., A global analysis of soil microbial biomass carbon, nitrogen, and phosphorus in terrestrial ecosystems, *Global Ecology and Biogeography*, 22:737–749 (2013).

6. P. G.Falkowski et al., The microbial engines that drive Earth's biogeochemical cycles, *Science*, 320: 1034–1039 (2008).

7. B. I. Jelen et al., The role of microbial electron transfer in the coevolution of the biosphere and geosphere, *Annual Reviews in Microbiology*, 70: 45–62 (2016).

8. B. E. Schirrmeister et al., Evolution of multicellularity coincided with increased diversification of cyanobacteria and the Great Oxidation Event, *Proceedings of the National Academy of Sciences*, 110: 1791–1796 (2013).

9. R. M. Hazen et al., Mineral evolution. *American Mineralogist*, 93: 1693–1720 (2008).

10. H. Jenny, Role of the plant factor in the pedogenic functions, *Ecology*, 39: 5–16 (1958).

11. R. DeWit and T. Bouvier, "Everything is everywhere, but, the environment selects": what did Bass Backing and Bekerinck really say? *Environmental Microbiology*, 8: 755–758 (2006).

12. K. M.Finstad et al., Microbial community structure and the persistence of cyanobacterial populations in salt crusts of the hyperarid Atacama Desert from genome-resolved metagenomics, *Frontiers in Microbiology*, 8: 1435 (2017).

13. M. Delgado-Baquerizo et al., Changes in belowground biodiversity during ecosystem development, *Proceedings of the National Academy of Sciences*, 116: 6891–6896 (2019).

14. W. E. Slessarev et al., Water balance creates a threshold in soil pH at the global scale, *Nature*, 540: 567–569 (2016).

15. M. Delgado-Baquerizo et al., A global atlas of the dominant bacteria found in soil, *Science*, 359: 320–325 (2018).

16. N. Fierer et al., Cross-biome metagenomic analysis of soil microbial communities and their functional attributes, *Proceedings of the National Academy of Sciences*, 109: 21390–21395 (2012).

17. R. Amundson et al., Soil diversity and land use in the United States, *Ecosystems*, 6: 470–482 (2004).

18. N. Fierer et al., Reconstructing the microbial diversity and function of pre-agricultural tallgrass prairie soils in the United States, *Science*, 342: 621–624 (2013).

19. K. S. Ramirez et al., Consistent effect of nitrogen amendments on soil microbial communities and processes across biomes, *Global Change Biology*, 18:1918–1927 (2012).

20. B. M. Hoover et al., Culture-independent discovery of the malacidins as calcium-dependent antibiotics with activity against multidrug-resistant Gram-positive pathogens, *Nature Microbiology*, 3: 415–422 (2018).

21. A. Crits-Christoph et al., Novel soil bacteria possess diverse genes for secondary metabolite biosynthesis, *Nature*, 558: 440–444 (2018).

22. W. R. Weider et al., Explicitly representing soil microbial processes in Earth System Models, *Global Biogeochemical Cycles*, 29: 1782–1800 (2015).

23. D. Bru et al., Determinants of the distribution of nitrogen-cycling microbial communities at the landscape scale, *The ISME Journal*, 5: 532–542 (2011).

24. M. Kennedy et al., Late Precambrian oxygenation: Inception of the clay mineral factory, *Science*, 311: 1446–1449 (2006).

25. A paleosol is a soil formed at the land surface that has been buried and become the equivalent of a biological fossil. A summary of Precambrian paleosol studies is found in: H. Beraldi-Campesi and G. J. Retallack, Terrestrial ecosystems in the Precambrian, chapter 3 in B. Weber et al. (eds), *Biological Soil Crusts: An Organizing Principle in Drylands*, Ecological Studies vol. 226, Springer, Cham. doi: 10.1007/978–3–319–30214-0_3.

26. S. G. Driese. Pedogenic translocation of Fe in modern and ancient Vertisols and implications for interpretations of the Hekport paleosols (2.25Ga), *Journal of Geology*, 1112: 543–560 (2004).

27. Kennedy et al., Late Precambrian oxygenation.

28. Ideas in this section come from an excellent review by Y. Lucas, The role of plants in controlling rates and products of weathering: Importance of biological pumping, *Annual Review of Earth and Planetary Science*, 29: 135–163 (2001).

29. Lucas, The role of plants.

30. L. A. Derry et al., Biological control of terrestrial silica cycling and export fluxes to watersheds, *Nature*, 433: 728–731 (2005).

31. E. G. Jobbágy and R. B. Jackson, . The distribution of soil nutrients with depth: Global patterns and the imprint of plants, *Biogeochemistry*, 53: 51–77 (2001).

32. H. He et al., Physiological and ecological significance of biomineralization in plants, *Trends in Plant Science*, 19: 166–174 (2014).

33. S. A. Quideau et al., Biogeochemical cycling of calcium and magnesium by ceanothus and chamise, *Soil Science Society of America Journal*, 63: 1880–1888 (1999).

34. F. di Castri and H. A. Mooney (eds), *Mediterranean Type Ecosystems*, Springer, New York, Heidelberg, Berlin (1973).

35. K. L. Moulton and R. A. Berner, Quantification of the effect of plants on weathering: studies in Iceland, *Geology*, 26: 895–898 (1998).

36. Kennedy et al., Late Precambrian oxygenation.

37. S. A. Ewing et al., A threshold in soil formation at Earth's arid–hyperarid transition, *Geochimica et Cosmochimica Acta*, 70: 5293–5322 (2006); S. A. Ewing et al., The rainfall limit of the N cycle on Earth, *Global Biogeochemical Cycles*, 21: GB3009 (2007); J. J. Owen et al., The sensitivity of bedrock erosion to precipitation, *Earth Surface Processes and Landforms*, 36: 117–135 (2011).

38. W. E. Dietrich and J. T. Perron, The search for a topographic signature of life, *Nature*, 439: 411–418 (2006).

39. D. H. Alban and E. C. Berry, Effects of earthworm invasion on morphology, carbon and nitrogen of a forest soil, *Applied Soil Ecology*, 1: 243–249 (1994).

40. S. Porder, G. E. Hilley, and O. A. Chadwick, Chemical weathering, mass loss, and dust inputs across a climate by time matrix in the Hawaiian Islands, *Earth and Planetary Science Letters*, 258: 414–427 (2007).

4 Field-Based Properties of Soils

1. T. C. Chamberlin, On Lord Kelvin's Address on the Age of the Earth as an Abode Fitted for Life, Smithsonian Institution Annual Report, Washington, DC, 223–246 (1899).

2. Soil Science Division Staff, *Soil Survey Manual*, C. Ditzler, K. Scheffe, and H. C. Monger (eds), USDA Handbook 18, Government Printing Office, Washington, DC (2017); www.nrcs.usda.gov/wps/portal/nrcs/detail/soils/ref/?cid=nrcs142p2_054262

3. See interesting discussions of the Russian influence in pedology in I. A. Krupenikov, *History of Soil Science: From Its Inception to the Present*, Russian Translation Series 98, A.A. Balkema Publishers, Brookfield, VT (1993); V. R. Vil'yams, *Dokuchaev's Role in the Development of Soil Science: Introduction to the Russian Chernozem (by V.V. Dokuchaev)*, Translated by the Israel Program for Scientific Translations, Jerusalem (1967).

4. Soil Survey Staff, *Soil Survey Manual*.

5. There are now several online Munsell color charts (which do not replace the physical version for the field but can be interactively explored).

6. The Keys to Soil Taxonomy is the key resource for classifying soil. However, it is recommended, for an introduction, that the reader access the downloadable "Illustrated Guide to Soil Taxonomy," which, in addition to containing most important information on soil classification, has numerous illustrations and photos relevant to fieldwork and soil identification: Soil Survey Staff, *Illustrated Guide to Soil Taxonomy*, U.S. Department of Agriculture, Natural Resources Conservation Service, National Soil Survey Center, Lincoln, Nebraska (2015).

7. U. Schwertmann, Relationship between iron oxides, soil color, and soil formation, pp. 51–69, in J. M. Bigham and E. J. Ciolkosz (eds), *Soil Color*, Soil Science Society of America Special Publication 31, Soil Science Society of America, Madison, WI (1993).

8. J. W. Harden, A quantitative index of soil development from field descriptions: examples from a chronosequence in central California, *Geoderma*, 28: 1–28 (1982).

9. Soil Survey Staff, *Soil Survey Manual*.

10. J. M. Oades, Soil organic matter and structural stability: Mechanisms and implications for management, *Plant and Soil*, 76: 319–337 (1984); J. M. Oades, The role of biology in the formation, stabilization, and degradation of soil structure, *Geoderma*, 56: 377–400 (1993).

11. E. M. White, Subsoil structure genesis: Theoretical consideration, *Soil Science*, 101:135–141 (1966); R. J. Southard and S. W. Buol, Subsoil blocky structure formation in some North Carolina Paleudults and Paleaqults, *Soil Science Society of America Journal*, 52: 1069–1076 (1988).

12. L. D. Whittig and P. Janitzky, Mechanisms of formation of sodium carbonate in soils: 1. Manifestations of biological conversions, *Journal of Soil Science*, 14: 322–333 (1963).

13. E. B. Alexander and W. D. Nettleton, Post-Mazama Natrargids in Dixie Valley, Nevada, *Soil Science Society of America Journal*, 41: 1210–1212 (1977).

14. M. Rieu and G. Sposito, Fractal fragmentation, soil porosity, and soil water properties. 1. Theory, *Soil Science Society of America Journal*, 55: 1231–1238 (1991).

15. Soil Survey Staff, *Soil Survey Manual*.

16. https://www.nrcs.usda.gov/wps/portal/nrcs/detail/soils/ref/?cid=nrcs142p2_054253#designations

5 Soil Biogeochemical Measurements and Data

1. The website for the NRCS database on soils is https://ncsslabdatamart.sc.egov.usda.gov.

2. Soil Survey Staff, *Soil Survey Laboratory Methods Manual*, Soil Survey Investigations Report No. 45, Natural Resource Conservation Service, USDA, Washington, DC (2011).

3. Soil Survey Staff, *Soil Survey Laboratory Methods Manual*, Soil Survey Investigations Report No. 42, Ver. 5.0, Natural Resource Conservation Service, USDA, Washington, DC (2014).

4. U. Schwertmann, Solubility and dissolution of iron oxides, *Plant and Soil*, 130: 1–25 (1991).

5. J. B. Dixon and S. B. Weed (eds), *Minerals in Soil Environments*, 2nd Edition, Soil Science Society of America, Madison, WI (1989).

6. For example, see F. Macht et al., Specific surface area of clay minerals: comparison between atomic force microscopy measurements and bulk-gas ($N2$) and liquid (EGME) adsorption methods, *Applied Clay Science*, 53: 20–26 (2011).

7. E. A. Bettis III et al., Last glacial loess in the conterminous USA, *Quaternary Science Reviews*, 22: 1907–1946 (2003).

8. W. M. Post et al., Soil carbon pools and world life zones, *Nature*, 298: 156–159 (1982).

9.   J. Z. Williams et al., The temperature dependence of feldspar dissolution determined using a coupled weathering-climate model for Holocene-aged loess soils, *Geoderma*, 156: 11–19 (2010).

10.   D. R. Muhs et al., Impact of climate and parent material on chemical weathering in loess-derive soils of the Mississippi River Valley, *Soil Science Society of America Journal*, 65: 1761–1777 (2001).

11.   Williams et al., The temperature dependence of feldspar dissolution.

6 Time and Soil Processes

1.   J. McPhee, *Basin and Range*, Farrar, Straus, Giroux, New York (1981).

2.   S. J. Gould, *Time's Arrow, Time's Cycle*, Harvard University Press, Cambridge, MA (1987); C. C. Albritton, Jr, *The Abyss of Time*, Freeman, Cooper, and Co., San Francisco, CA (1980); W. B. N. Berry, *Growth of a Prehistoric Time Scale*, W.H. Freeman and Company, San Francisco, CA (1968).

3.   W. B. Bull, *Geomorphic Response to Climate Change*, Oxford University Press, New York (1991), p. 316.

4.   L. A. Hug et al., A new view of the tree of life, *Nature Microbiology*, doi:10.1038 (2016).

5.   W. B. Whitman et al., Prokaryotes: The unseen majority, *Proceedings of the National Academy of Sciences*, 95: 6578–6583 (1998).

6.   D. Christian, *Origin Story. A Big History of Everything*, Little Brown Spark, New York (2018).

7.   For a recent discussion, see R. A. Berner, The rise of plants and their effect on weathering and atmospheric CO_2, *Science*, 276: 544–545; M. Kennedy et al., Late Precambrian oxygenation: inception of the clay mineral factory, *Science*, 311: 1446–1449 (2006).

8.   T. E. Cerling et al., Global vegetation change through the Miocene/Pliocene boundary, *Nature*, 389:153–158 (1997); G. Rettallack, Cenozoic paleoclimate on land in North America, *The Journal of Geology*, 115: 271–294 (2007).

9.   The long-term stability of ancient landscapes (for example, fluvial deposits or terraces) is certainly subject to debate, and in an absolutely strict sense, most landscapes have been altered to some degree since their original formation. However, some retain morphological characteristics that suggest that erosional alteration has been "minor," allowing us to glimpse the result of soil formation over vast expanses of time. Two well documented examples are Holocene to Pliocene volcanic flows in Hawaii (e.g. P. M. Vitousek et al., Soil and ecosystem development across the Hawaiian Islands, *GSA Today*, 7(9):1–8 (1997) and Holocene to Pliocene river terraces in California (J. W. Harden, Soils developed in granitic alluvium near Merced, CA, pp. 1–65, in: J. W. Harden (ed), A Series of Soil Chronosequences in the Western United States, U.S. Geological Survey Bulletin 1590-A (1987); A. F. White et al., Chemical weathering rates of a soil chronosequence on granitic alluvium. I. Quantification of mineralogical and surface area changes and calculation of primary silicate reaction rates, *Geochimica et Cosmochimica Acta*, 60:2533–2550 (1996).

10.   P. J. Crutzen, Geology of mankind, *Nature*, 415: 23 (2002).

11. G. J. Retallack, *Soils of the Past: An Introduction to Paleopedology*, 2nd Edition, Wiley, Hoboken, NJ (2001).

12. Y. Wang, R. Amundson, and S. Trumbore, Radiocarbon dating of soil organic matter, *Quaternary Research*, 45: 282–288 (1996).

13. R. Amundson et al., Factors and processes governing the carbon-14 content of carbonate in desert soils, *Earth and Planetary Science Letters*, 125: 285–405 (1994).

14. D. E. Granger and C. S. Riebe, Cosmogenic Nuclides in Weathering and Erosion, in: H. D. Holland and K. K. Turekian (eds), *Treatise on Geochemistry*, Volume 5: Surface and Ground Water, Weathering, and Soils, 2nd Edition, Elsevier, Oxford, 401–436 (2014).

15. E. J. Oerter et al., Pedothem carbonates reveal anomalous North American atmospheric circulation 70,000–55,000 years ago, *Proceedings of the National Academy of Sciences*, 113: 919–924 (2016).

16. D. J. Merritts et al., The mass balance of soil evolution on late Quaternary marine terraces, northern California, *Geological Society of America Bulletin*, 104: 1456–1470 (1992); G. H. Brimhall et al., Deformational mass transport and invasive processes in soil evolution, *Science*, 255: 695–702 (1992); D. J. Merritts et al., Rates and processes of soil evolution on uplifted marine terraces, northern California, *Geoderma*, 51:241–275 (1991); C. A. Masillo et al., Weathering controls on mechanisms on carbon storage in grassland soils, *Global Biogeochemical Cycles*, doi: 10.1029/2004GB002219 (2004).

17. M. C. Jungers et al., Active erosion-deposition cycles in the hyperarid Atacama Desert of northern Chile, *Earth and Planetary Science Letters*, 371–372: 125–133 (2013); R. S. Anderson et al., Explicit treatment of inheritance in dating depositional surfaces using in situ Be-10 and Al-26, *Geology*, 24:47–51 (1996).

18. L. A. Perg et al., Use of a new 1Be and 26Al inventory method to date marine terraces, Santa Cruz, California, USA, *Geology*, 29:879–882 (2001).

19. D. R. Montgomery, *The Rocks Don't Lie: A Geologist Investigates Noah's Flood*, W.W. Norton & Company (2012).

7 The Soil Carbon Cycle

1. C. Darwin, *The Formation of Vegetable Mould Through the Action of Worms, with Observations on their Habits*, John Murray, London (1881).

2. J. M. Oades, An introduction to organic matter in mineral soils, in: J. B. Dixon and S. B. Weed, *Minerals in Soil Environments*, 2nd Edition, Soil Science Society of America, Madison, WI (1989).

3. e.g. L. H. P. Jones and K. A. Handreck, Silica in soils, plants, and animals, *Advances in Agronomy*, 19: 107–149 (1967).

4. A. H. Jahren, M. L. Gabel, and R. Amundson, Biomineralization in seeds: development trends in isotopic signatures of hackberry, *Palaeogeography, Palaeoclimatology, and Palaeoecology*, 138: 259–269 (1998).

5. L. R. Drees et al., Silica in soils: quartz and disordered silica polymorphs, in: Dixon and Weed (eds), *Minerals in Soil Environments*.

6. W. J. Parton et al., Analysis of factors controlling soil organic matter levels in Great Plains grasslands, *Soil Science Society of America Journal*, 51:1173–1179 (1987).

7. R. Sutton and G. Sposito, Molecular structure in soil humic substances: The new view, *Environmental Science and Technology*, 39: 9009–9015 (2005).

8. J. Lehmann and M. Kleber, The contentious nature of soil organic matter, *Nature*, 528: 60–68 (2015).

9. I. Bisutti, I. Hilke, and M. Raessler, Determination of total organic carbon: an overview of current methods, *Trends in Analytical Chemistry*, 23: 716–726 (2004).

10. R. Amundson et al., Global patterns of the isotopic composition of soil and plant nitrogen, *Global Biogeochemical Cycles*, 17: doi: 10.1029/2002GB001903 (2003); W. T. Baisden et al., A multi-isotope C and N modeling analysis of soil organic matter turnover and transport as a function of soil depth in a California annual grassland, *Global Biogeochemical Cycles*, 16: doi: 1029/2001GB001823 (2002).

11. R. F. Stallard, Terrestrial sedimentation and the carbon cycle: Coupling weathering and erosion to the carbon cycle, *Global Biogeochemical Cycles*, 12: 231–257 (1998).

12. P. J. Hanson et al, Separating root and soil microbial contributions to soil respiration: A review of methods and observations, *Biogeochemistry*, 48, 115–146 (2000).

13. R. Amundson and W. T. Baisden, Stable isotope tracers and mathematical models in soil organic matter studies, pp. 117–137, in: O. E. Sala et al. (eds), *Methods in Ecosystem Science*, Springer, New York (2000).

14. J. Kaste et al., Short term soil mixing quantified with fallout radionuclides, *Geology*, 35: 243–246 (2007).

15. M. Jagercikova et al., Vertical distributions of 137Cs in soils: a meta-analysis, *Journal of Soils and Sediments*, 15: 81–95 (2015).

16. https://daac.ornl.gov/SOILS/guides/HWSD.html

17. M. A. Bradford et al., Managing uncertainty in soil carbon feedbacks to climate change, *Nature Climate Change*, 6: 751–758 (2016); T. W. Crowther et al., Quantifying soil carbon losses in response to warming, *Nature*, 540:104–108 (2016); E. A. Davidson and I. A. Janssens, Temperature sensitivity of soil carbon decomposition and feedbacks to climate change, *Nature*, 440: 165–173 (2006).

18. http://fluxnet.fluxdata.org

19. J. Sanderman et al., Application of eddy covariance measurements to the temperature dependence of soil organic matter mean residence time, *Global Biogeochemical Cycles*, 17:doi:10.1029/2001GB001833 (2003)

20. J. Harden et al., Dynamics of soil carbon during deglaciation of the Laurentide ice sheet, *Science*, 258: 1921–1924 (1992).

21. S. Trumbore et al., Rapid exchange between soil carbon and atmospheric carbon dioxide driven by temperature change, *Science*, 272: 393–396 (1996).

22. C. D. Koven et al., Higher climatological temperature sensitivity of soil carbon in cold than warm climates, *Nature Climate Change*, 7: 817–824 (2017).

23. Amundson and Baisden, Stable isotope tracers.

24. Amundson and Baisden, Stable isotope tracers.

25. S. E. Trumbore et al.,Radiocarbon nomenclature, theory, models, and interpretation: Measuring age, determining cycling rates, and tracing source pools, in: E. Schuur, E. Druffel, and S. Trumbore (eds), *Radiocarbon and Climate Change*, Springer, Switzerland (2016).

26. M. Tifafi et al., The use of radiocarbon 14C to constrain carbon dynamics in the soil module of the land surface model ORCHIDEE (SVN r5165), *Geoscience Model Development*, 11: 4711–4726 (2018).

27. Baisden et al., A multi-isotope C and N modeling analysis.

28. J. A. Mathieu et al., Deep soil carbon dynamics are driven more by soil type than by climate: a world-wide meta-analysis of radiocarbon profiles, *Global Change Biology*, 21: 4278–4292 (2015).

29. S. A. Ewing et al., Role of large-scale soil structure in organic carbon turnover: Evidence from California grassland soils, *Journal of Geophysical Research*, 111, doi: 10.1029/2006JG000174 (2006).

30. R. Amundson et al., The isotopic composition of soil and soil-respired CO_2, *Geoderma*, 82: 83–114 (1998).

31. P. Moldrup et al., Modeling diffusion and reaction in soils IX. The Buckingham-Burdine-Campbell equation for gas diffusivity in undisturbed soils, *Soil Science*, 164: 542–551 (1999).

32. B. Bond-Lamberty et al., Globally rising soil heterotrophic respiration over recent decades, *Nature*, 560: 80–3 (2018).

33. J. N. Butler, Carbon Dioxide Equilibria and Their Applications, Addison-Wesley Publishing Company, Inc., Reading, MA (1982).

34. G. Retallack, Pedogenic carbon proxies for amount and seasonality of precipitation in paleosols, *Geology*, 33: 333–336 (2004).

35. T. E. Cerling, The stable isotopic composition of modern soil carbonate and its relationship to climate, *Earth and Planetary Science Letters*, 71: 229–240 (1984).

36. Amundson et al., The isotopic composition of soil.

37. Z. Sharp, *Principles of Stable Isotope Geochemistry*, 2nd Edition, doi: https://doi.org/10.25844/h9q1-0p82 (2017).

38. Sharp, *Principles of Stable Isotope Geochemistry*.

39. A. Ebeling et al., Relict soil evidence for profound Quaternary aridification of the Atacama Desert, Chile, *Geoderma*, 267: 196–206 (2016).

40. See the online assemblage of temperature-dependent isotopes fractionation factors: www2.ggl.ulaval.ca/cgi-bin/alphadelta/alphadelta.cgi

41. Global soil C storage in atmosphere (800 Gt C)/Annual inputs to soil (60 Gt C y^{-1}) =13.3 y.

8 Chemical and Physical Processes in Soils

1. W. Stumm and J. J. Morgan, Aquatic chemistry, in: *Chemical Equilibria and Rates in Natural Waters*, 3rd Edition, John Wiley and Sons, New York (1996).

2. Examples of soil water chemistry papers that complement corresponding papers on the soil solid-phase chemistry include: A. F. White et al., Chemical weathering rates of a soil chronosequence on granitic alluvium: III. Hydrochemical evolution and contemporary solute fluxes and rates, *Geochimica et Cosmochimica Acta*, 69: 1975–1996 (2005); A. F. White et al., Chemical weathering of a marine terrace chronosequence, Santa Cruz, California. Part II: Solute profiles, gradients and the

comparisons of contemporary and long-term weathering rates, *Geochimica et Cosmochimica Acta*, 73: 2769–2803 (2009).

3. G. H. Brimhall and W. E. Dietrich, Constitutive mass balance relations between chemical-composition, volume, density, porosity, and strain in metasomatic hydro-chemical systems – results on weathering and pedogenesis, *Geochimica et Cosmochimica Acta*, 51: 567–587 (1987); see also, for discussions including dust and other derivations, S. A. Ewing et al., A threshold in soil formation at Earth's arid-hyperarid transition, *Geochimica et Cosmochimica Acta*, 70: 5293–5322 (2006).

4. L. B. Railsback, An earth scientist's periodic table of the elements and their ions, *Geology*, 31: 737–740 (2003).

5. L. P. Gromet and L. T. Silver, Rare earth element distributions among minerals in a granodioriate and the petrogenetic implications, *Geochimica et Cosmochimica Acta*, 47: 925–939 (1983).

6. A. Ebeling et al., Relict soil evidence for profound quaternary aridification of the Atacama Desert, Chile,*Geoderma*, 267: 196–206 (2016).

7. A. F. White et al., Chemical weathering of a marine terrace chronosequence, Santa Cruz, California. I: Interpreting rates and controls based on soil concentration-depth profiles, *Geochimica et Cosmochimica Acta*, 72: 36–68 (2008); A. F. White et al., Chemical weathering of a marine terrace chronosequence, Santa Cruz, California. Part II: Solute profiles, gradients and the comparisons of contemporary and long-term weathering rates, *Geochimica et Cosmochimica Acta*, 73: 2769–2803 (2009); K. Maher et al., The role of reaction affinity and secondary minerals in regulating chemical weathering rates at the Santa Cruz soil chronosequence, California, *Geochimica et Cosmochimica Acta*, 73: 2804–2831 (2009); J. More et al., Shifting microbial community structure across a marine terrace grassland chronosequence, Santa Cruz, California, *Soil Biology and Biochemistry*, 42: 21–31 (2010); and a number of others.

8. L. A. Perg et al., Use of a new 1Be and 26Al inventory method to date marine terraces, Santa Cruz, California, USA, *Geology*, 29: 879–882 (2001).

9. F. White et al., Chemical weathering of a marine terrace chronosequence, Santa Cruz, California I: Interpreting rates and controls based on soil concentration-depth profiles, *Geochimica et Cosmochimica Acta*, 72: 36–68 (2008).

10. White et al., Chemical weathering.

11. A. Navarre-Sitchler and G. Thyne, Effects of carbon dioxide on mineral weathering rates at earth surface conditions, *Chemical Geology*, 243: 53–63 (2007); R. A. Berner, The rise of land plants and their effect on weathering and atmospheric CO_2, *Science*, 276: 544–546 (1997).

12. R. A. Berner and Z. Kothavala, GEOCARB III: A revised model of atmospheric CO2 over Phanerozoic time, *American Journal of Science*, 301: 182–204 (2001).

13. A. F. White and A. E. Blum, Effects of climate on chemical-weathering in watersheds, *Geochimica et Cosmochimica Acta*, 59: 1729–1747 (1995).

14. A. T. Evan et al., The past, present and future of African dust, *Nature*, 531 (7595): 493–497 (2016).

15. O. A. Chadwick et al., Changing sources of nutrients during four million years of ecosystem development, *Nature*, 397: 491–497 (1999).

16. L. D. McFadden et al., Influences of eolian and pedogenic processes on the evolution and origin of desert pavements, *Geology*, 15: 504–508 (1987).

17.   S. A. Ewing et al., A threshold in soil formation at Earth's arid-hyperarid transition, *Geochimica et Cosmochimica Acta*, 70: 5293–5322 (2006); S. A. Ewing et al., The rainfall limit of the nitrogen cycle on Earth, *Global Biogeochemical Cycles*, 21: GB3009 (2007); B. Sutter et al., Terrestrial analogs for interpretation of infrared spectra from the Martian surface and subsurface: sulfate, nitrate, carbonate, and phyllosilicate-bearing Atacama Desert soils, *Journal of Geophysical Research-Biogeosciences*, 112: GO4S10 (2007); S. A. Ewing et al., Non-biological fractionation of stable Ca isotopes in soils of the Atacama Desert, Chile, *Geochimica et Cosmochimica Acta*, 72: 1096–1110 (2008); S. A. Ewing et al., Changes in the soil C cycle at the arid-hyperarid transition in the Atacama Desert, *Journal of Geophysical Research-Biogeosciences*, 113: G02S90 (2008); R. Amundson et al., The stable isotope composition of halite and sulfate of hyperarid soils and its relation to aqueous transport, *Geochimica et Cosmochimica Acta*, 99: 271–286 (2012); M. C. Jungers et al., Active erosion-deposition cycles in the hyperarid Atacama Desert of northern Chile, *Earth and Planetary Science Letters*, 371: 125–133 (2013).
18.   R. Amundson et al., On the in situ aqueous alteration of soils on Mars, *Geochimica et Cosmochimica Acta*, 72: 3845–3864 (2008).
19.   Brimhall and Dietrich, Constitutive mass balance relations.

9 Soil Processes on Sloping Landscapes

1.   H. Jenny, *Factors of Soil Formation*, McGraw-Hill Book Co., New York (1941).
2.   *Glossary of Soil Science Terms*, Soil Science Society of America, Madison, WI (2008).
3.   G. K. Gilbert, Report on the Geology of the Henry Mountains, Dept. of the Interior, US Govt. Printing Office, Washington, DC (1877).
4.   J. T. Perron, Climate and the pace of erosional landscape evolution, *Annual Reviews in Earth and Planetary Science*, 45: 561–591 (2017).
5.   A. Heimsath et al., The illusion of diffusion: field evidence for depth-dependent sediment transport, *Geology*, 33: 949–952 (2005).
6.   Gilbert, Report on the Geology of the Henry Mountains; G. K. Gilbert, The convexity of hilltops, *Journal of Geology*, 17: 344–350 (1909).
7.   A. M. Heimsath et al., The soil production function and landscape equilibrium, *Nature*, 388: 358–361 (1997).
8.   R. Amundson et al., Hillslope soils and vegetation, *Geomorphology*, 234: 122–132, (2015); Perron, Climate.
9.   Perron, Climate.
10.   K. Yoo et al., Integration of geochemical mass balance with sediment transport to calculate rates of soil chemical weathering and transport on hillslopes, *Journal of Geophysical Research*, 112: F02013 (2007).
11.   K. Yoo et al., Erosion of upland hillslope soil organic carbon: Coupling field measurements with a sediment transport model, *Global Biogeochemical Cycles*, 19: GB3003 (2005).
12.   C. Rasmussen et al., Strong climate and tectonic control on plagioclase weathering in granitic terrain, *Earth and Planetary Science Letters*, 301: 521–530 (2011).

13. S. Porder et al., Ground-based and remotely sensed nutrient availability across a tropical landscape, *Proceedings of the National Academy of Sciences*, 102: 10 909–10 912 (2005).

14. A. Eger et al., Does soil erosion rejuvenate the soil phosphorus inventory? *Geoderma*, 332: 45–59 (2018).

15. P. D'Odorico, Possible bistable evolution of soil thickness, *Journal of Geophysical Research*, 105: 25 927–25 935 (2000).

16. Amundson et al., Hillslope soils and vegetation.

17. Gilbert, Report on the Geology of the Henry Mountains.

18. Yoo et al., Erosion.

10 Humans and Soil Biogeochemistry

1. E. C. Ellis et al., Used planet: A global history, *Proceedings of the National Academy of Sciences*, 110: 7978–7985 (2013).

2. R. Amundson et al., Soil diversity and land use in the United States, *Ecosystems*, 6: 470–482 (2003)

3. Y. Guo et al., Taxonomic structure, distribution, and abundance of soils in the USA, *Soil Science Society of America Journal*, 67: 1507–1516 (2003).

4. Amundson et al., Soil diversity.

5. P. H. Kahn Jr, Children's affiliations with nature: structure, development, and the problem of environmental generational amnesia, pp. 93–116 in: P. H. Kahn Jr. and S. R. Kellert (eds), *Children and Nature: Psychological, Sociocultural, and Evolutionary Investigations*, MIT Press (2002).

6. Amundson et al., Soil diversity.

7. P. R. Ehrlich and E. O. Wilson, Biodiversity studies: science and policy, *Science*, 253: 758–762 (1991).

8. R. Amundson and H. Jenny, On a state factor model of ecosystems, *Bioscience*, 47: 536–543 (1997).

9. L. B. Guo and R. M. Gifford, Soil carbon stocks and land use change: A meta analysis, *Global Change Biology*, 8: 345–360 (2002).

10. R. Amundson and H. Jenny, The place of humans in the state factor theory of ecosystems and their soils, *Soil Science*, 151: 99–109 (1991); R. Amundson and H. Jenny, On a state factor model of ecosystems, *Bioscience*, 47: 536–543 (1997).

11. H. Jenny, Soil Fertility Losses under Missouri Conditions, University of Missouri Agricultural Experiment Station Bulletin 324 (1933).

12. R. Amundson et al., Soil and human security in the 21st century, *Science*, 348: 1261071.

13. J. Sanderman et al., Soil carbon debt of 12,000 years of human land use, *Proceedings of the National Academy of Sciences*, 114: 9575–9580.

14. C. Le Quéré et al., Global carbon budget 2018, *Earth System Science Data*, 10: 2141–2194 (2018).

15. D. Bossio et al., The role of soil carbon in natural climate solutions, *Nature Sustainability*, doi: 10.1038/s41893-020-0491z

16. M. Wiesmeier et al., Projected loss of soil organic carbon in temperate agricultural soils in the 21st century: Effects of climate change and carbon input trends, *Scientific Reports*, 6, 32525 (2016).

17. K. Arroues, NRCS, Hanford, CA.

18. Y. Watanabe et al., Ammonium ion exchange of synthetic zeolites: The effect of their open-window sizes, pore structures, and cation exchange capacities, *Separation Science and Technology*, 39: 2091–2104 (2005).

19. C. D. Koven et al., Higher climatological temperature sensitivity on soil carbon in cold than warm climates, *Nature Climate Change* (2017). doi: 10.1038/ NCLIMATE3421

20. W. M. Post et al., Soil carbon pools and world life zones, *Nature*, 298: 156–159 (1982).

21. B. Bond-Lamberty et al., Globally rising soil heterotrophic respiration over recent decades, *Nature*, 560: 80–83 (2018).

22. R. Amundson, The carbon budget in soils, *Annual Reviews of Earth and Planetary Science*, 29: 535–562 (2001).

23. W. E. Dietrich et al., Geomorphic transport laws for predicting the form and evolution of landscapes, pp. 103–132 in: P. Wilcock and R. Iverson (eds), *Prediction in Geomorphology*, AGU Geophysical Monograph Series, V. 135 (2003).

24. P. Borrelli et al., An assessment of the global impact of 21st century land use change on soil erosion, *Nature Communications* (2013). doi: 10.1038/s41467-017-02142-7

25. C. Di Stefano and V. Ferro, Establishing soil loss tolerance: An overview, *Journal of Agricultural Engineering*, XLVII: 127–133 (2016).

26. D. R. Montgomery, Soil erosion and agricultural sustainability, *Proceedings of the National Academy of Sciences*, 104, 13 268–13 272 (2007).

27. N. A. Jelinski et al., Meteoric beryllium-10 as a tracer of erosion due to postsettlement land use in west-central Minnesota, USA, *Journal of Geophysical Research: Earth Surface*, doi:10.1029/2018JF004720

28. N. A. Jelinski and K. Yoo, The distribution and genesis of eroded phase soils in the conterminous United States, *Geoderma*, 279: 149–164 (2016).

29. S. D. Prince et al., Net primary production of US Midwest croplands from harvest yield data, *Ecological Applications*, 11: 1194–1205 (2001).

30. J. M. Briggs and A. K. Knapp, Annual variability in primary production in tallgrass prairie: climate, soil moisture, topographic position, and fire as determinants of aboveground biomass, *American Journal of Botany*, 82: 1024–1030 (1995).

31. R. F. Stallard, Terrestrial sedimentation and the carbon cycle: coupling weathering and erosion to carbon burial, *Global Biogeochemical Cycles*, 12: 231–257 (1998).

32. S. A. Ewing et al., Role of large-scale soil structure in organic carbon turnover: Evidence from California grassland soils, *Journal of Geophysical Research*, 111 (2006). doi: 10.1029/2006JG000174

33. Stallard, Terrestrial sedimentation.

34. Stallard, Terrestrial sedimentation.

35. Z. Wang et al., Human-induced erosion has offset one-third of carbon emissions from land cover change, *Nature Climate Change* (2017). doi: 10.1038/NCLIMATE3263

36. J. N. Qinton et al., The impact of agriculture soil erosion on biogeochemical cycling, *Nature Geoscience* (2010). https://doi.org/10.1038/ngeo838

37. Amundson and Jenny, The place of humans; Amundson and Jenny, On a state factor model of ecosystems.

38. R. Amundson and L. Biardeau, Soil carbon sequestration is an elusive climate mitigation tool, *Proceedings of the National Academy of Sciences*, 115: 11 652–11 656 (2018).

Afterword

1 R. Amundson, The carbon budget in soils, *Annual Reviews of Earth and Planetary Sciences*, 29: 535–562 (2001).

2 R. Amundson, Soil Formation, chapter 5, pp. 1–35, in H. D. Holland and K. K. Turekian (eds), *Treatise of Geochemistry*, Elsevier, 2003.

3 R. Amundson, Soil Formation, chapter 7, pp. 1–26, in H. D. Holland and K. K. Turekian (eds), *Treatise of Geochemistry*, 2nd Edition, Elsevier, 2014.

Index

Printed in the United States
by Baker & Taylor Publisher Services